GROUNDWATER POTENTIAL MAPPING

A Practical Approach

THE AUTHORS

Dr. Deepak Patle is presently working as a Research Associate at ICAR's NAHEP-CAAST-CSDA Project in College of Agricultural Engineering, Jawaharlal Nehru Krishi Vishwa Vidyalaya (JNKVV) Jabalpur. He has completed his Ph.D. (SWE) from JNKVV Jabalpur in 2022; M.Tech. (SWE) from JNKVV Jabalpur in 2018 and B.Tech. (Agricultural Engineering) from MGCGV, Chitrakoot in 2015. He has qualified ICAR's ASRB-NET in 2019. He has been recipient of CSDA Fellowship (NAHEP-CAAST-CSDA, JNKVV Jabalpur) and also awarded with M.P. Young Scientist Fellowship from MPCST Bhopal. He has published more than 20 research articles in NAAS rated journals (Nationals and Internationals). He has also presented technical posters, conference papers, abstracts and book chapters in many national and international seminar, webinar, symposium, and conferences. Currently, His peer review experience extends to many renowned national and international journals with a high impact factor.

Dr. M.K. Awasthi is currently working as a Chief Scientist in AICRP on Irrigation Water Management at College of Agricultural Engineering, Jawaharlal Nehru Krishi Vishwa Vidyalaya (JNKVV) Jabalpur. He has 39-year-old strong background in agricultural engineering, soil & water conservation, groundwater management, and natural resource management. He has published more than 100 research articles including book chapters, technical bulletin/ manuals, and short communications in different renowned national and international scientific journals.

Dr. Sourabh Nema is working as Scientist – 'C' at National Institute of Hydrology, North West Regional Centre in Jodhpur (Rajasthan). He has more than 15 year's experience in application of Remote Sensing in Natural Resource Management. His expertise in Hydrological Modeling, Climate change, Environment Modeling, Artificial intelligence, and Ground Water Management. He has published many articles, book chapters, technical bulletin/ manuals, and short communications in different renowned national and international scientific journals.

Dr. R.K. Nema is currently working as a Professor & Head in the Department of Soil & Water Engineering at College of Agricultural Engineering, Jawaharlal Nehru Krishi Vishwa Vidyalaya (JNKVV) Jabalpur. He has 39-year-old professional experience in agricultural engineering, soil & water conservation, Command Area, Remote Sensing, and Groundwater Management. He has published more than 70 research articles including book chapters, technical bulletin/ manuals, and short communications in different renowned national and international journals.

GROUNDWATER POTENTIAL MAPPING

A Practical Approach

Dr. Deepak Patle

Research Associate
NAHEP-CAAST-CSDA
Department of Soil and Water Engineering
College of Agricultural Engineering, JNKVV Jabalpur 482 004 (MP)

Dr. M. K. Awasthi

Chief Scientist
AICRP on Irrigation Water Management
Department of Soil and Water Engineering
College of Agricultural Engineering, JNKVV Jabalpur 482 004 (MP)

Dr. Sourabh Nema

Scientist – 'C'
National Institute of Hydrology,
North West Regional Centre, Jodhpur 342 003 (Rajasthan)

Dr. R. K. Nema

Professor and Head
Department of Soil and Water Engineering
College of Agricultural Engineering, JNKVV Jabalpur 482 004 (MP)

2025

Daya Publishing House®

A Division of

Astral International Pvt. Ltd.

New Delhi – 110 002

ISBN: 9789359199580

Published by : **Daya Publishing House®**
A Division of
Astral International Pvt. Ltd.
– ISO 9001:2015 Certified Company –
4736/23, Ground Floor, Ansari Road,
Darya Ganj, New Delhi-110 002
Ph. 011-43549197, 23278134
e-mail: info@astralint.com
Website: www.astralint.com

PREFACE

The book entitled "**Groundwater Potential Mapping: A Practical Approach**" provides a detailed process for groundwater potential mapping on River Basin level. The manual covers the practical aspects of groundwater potential mapping, such as watershed delineation, data collection, data analysis, decision making criterion approach, map interpretation, and validation using remote sensing, GIS, and statistical modelling.

The groundwater potential maps have been developed through comprehensive data on different ground water influencing factors like geology, geomorphology, lineament density, land Use/ land cover, soil, slope, drainage density and rainfall using Remote Sensing technology and Geographic Information System (GIS).

The practical manual is intended to be a valuable resource for planners, engineers, and other professionals involved in the development and management of groundwater resources on River Basin Level. It is also a valuable resource for the general public who is interested in learning more about the groundwater resource mapping through RS & GIS. The document will help the users especially the hydrologists, hydrogeologists and engineers to make use of these maps effectively on the ground for siting wells and for constructing recharge structures.

We would like to sincerely thank the AICRP on Irrigation Water Management (IWM) for providing their valuable suggestions for this practical manual.

We extend our heartfelt gratitude to all those who have contributed to this project. The technical support provided by the Indian Council of

Agricultural Research (ICAR) - National Agricultural Higher Education Project (NAHEP) - Centre for Advanced Agricultural Science and Technology (CAAST) through Jawaharlal Nehru Krishi Vishwa Vidyalaya is highly acknowledge without which the study could not be possible.

Deepak Patle
M. K. Awasthi
Sourabh Nema
R. K. Nema

CONTENTS

BASE MAP PREPARATION OF RIVER BASIN

Study areas can be watersheds, villages, Talukas, Tehsils, blocks, districts, states, or any other geographical area. There is no limitation on the size of the study area, which can be depends on the type of research. Different type of maps such as villages, Talukas, Tehsils, blocks, districts, states or entire nation, can be developed with the help of Survey of India (SOI) Toposheets available on different scale. In this research, one of the river basins of Madhya Pradesh was selected as study area. The process of development of Ken River Basin map is given as below:

Open the latest LTR version of QGIS and click on **Plugins** menu. Go to **Manage and Install Plugins** option and install the **SRTM-Downloader** plugin.

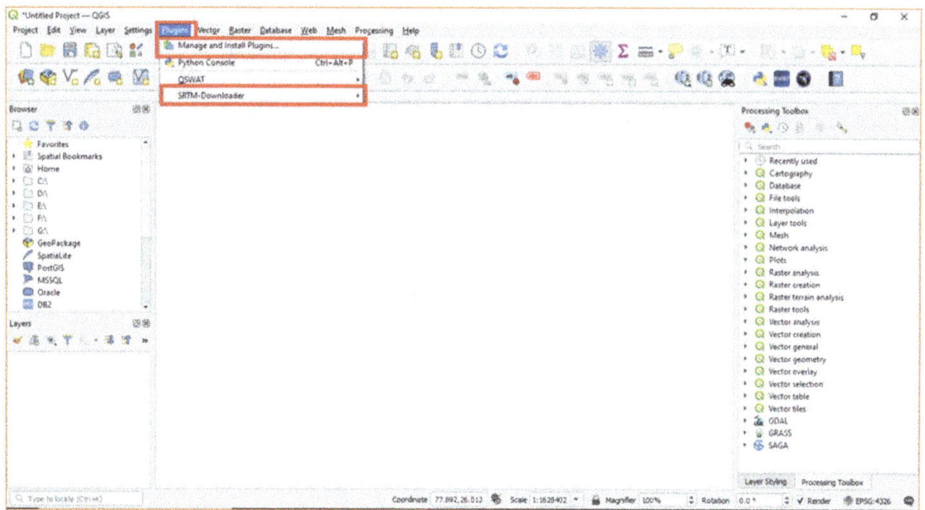

Click on the **SRTM Downloader.**

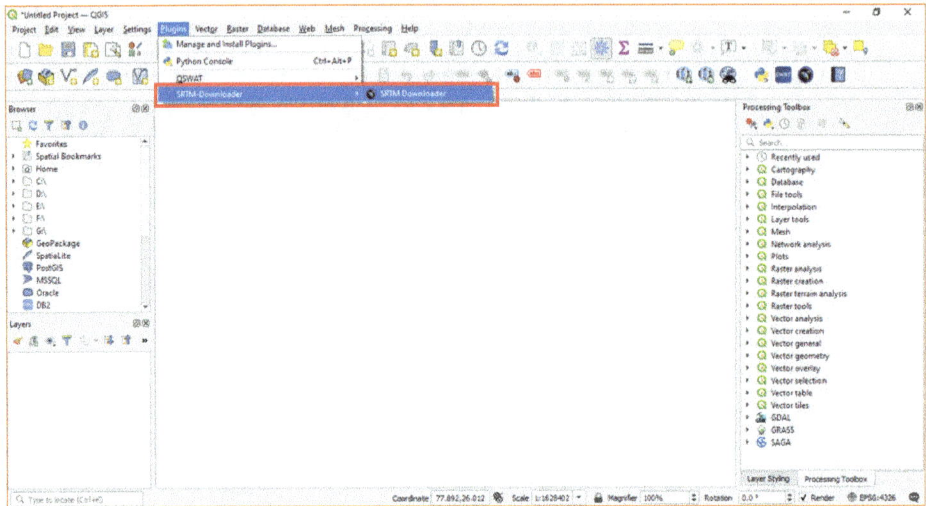

After clicked on **SRTM Downloader** option, the dialog box will open. Fill the **North-South** and **East-West** values of extent. Before entering the extent value, we need to know about the min and max extent (lat-long) for the River Basin.

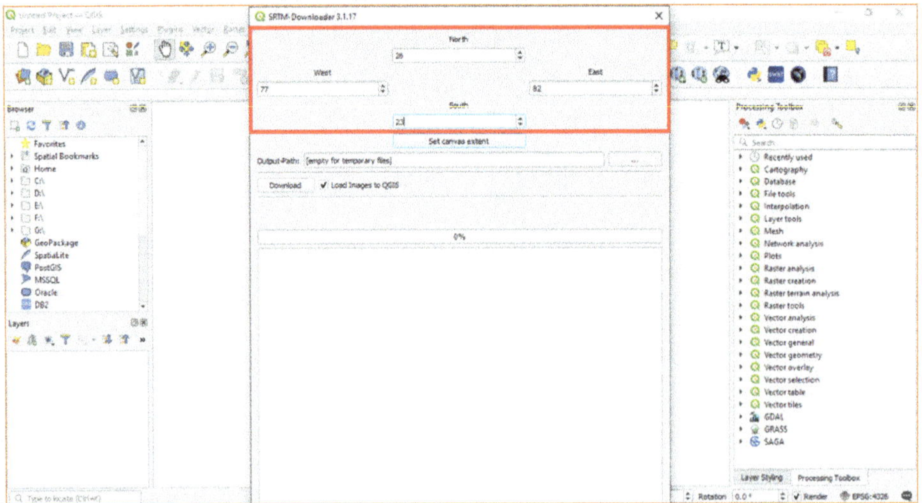

After clicked on **Set Canvas Extent** option, set the output path for temporary files of DEM. Just click on **Download** option then a dialog box will open with asking **Username** and **Password** of account in earth data. For this, it is required to have an account on earth data portal (https://urs.earthdata.nasa.gov/users/new). After filling the details, hit the **OK** button.

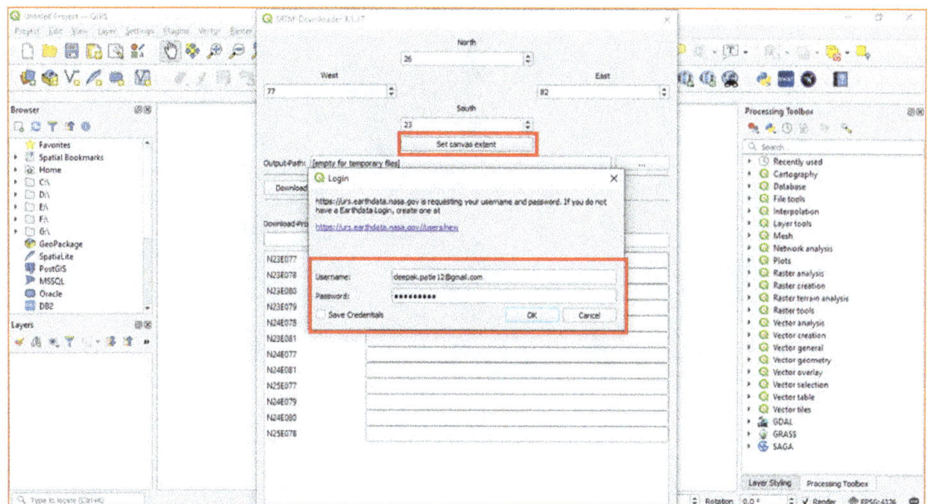

After that, the **Download-Progress** option will display the percentage of downloading tiles of DEM.

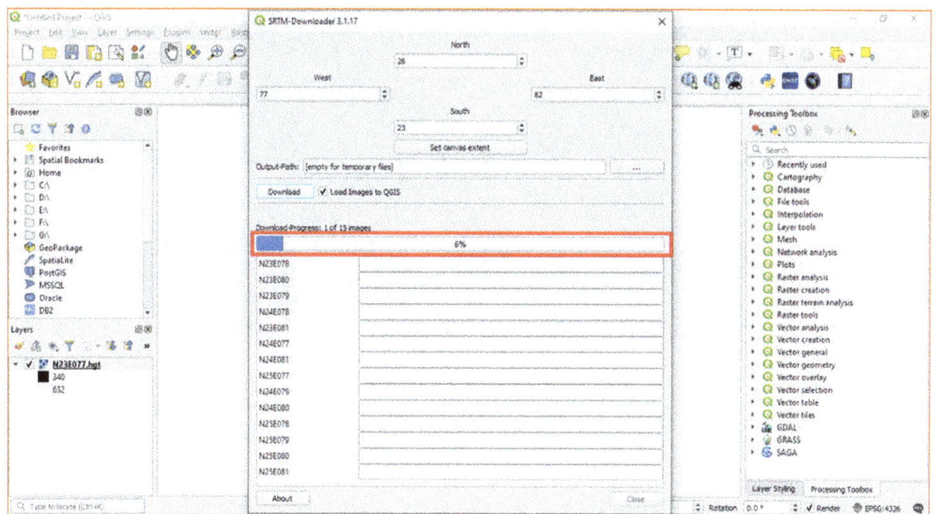

When the DEM tiles download, they appear in the **Layers** panel. The **Layers** panel is available in left corner of the QGIS interface.

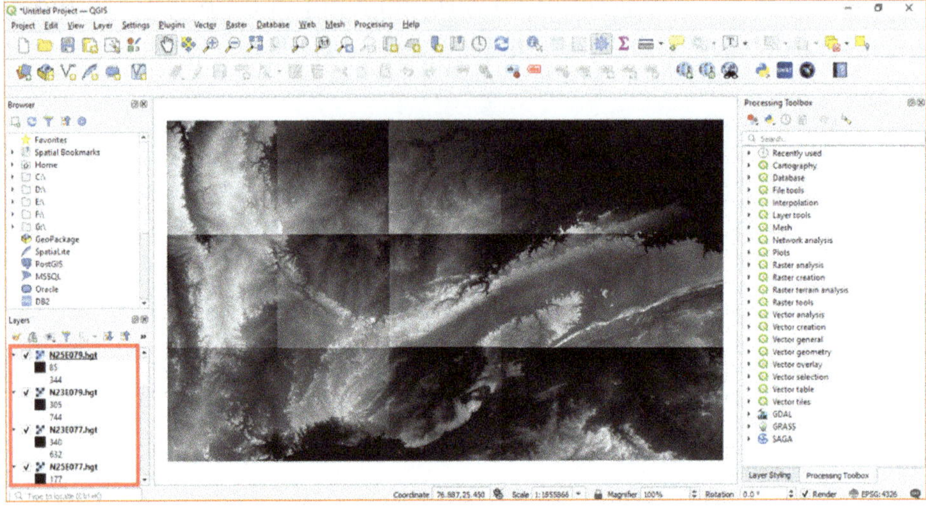

Now, all the tiles of DEM need to merge. Click on **Raster** menu, then go to **Miscellaneous** option, and click on the **Merge** option.

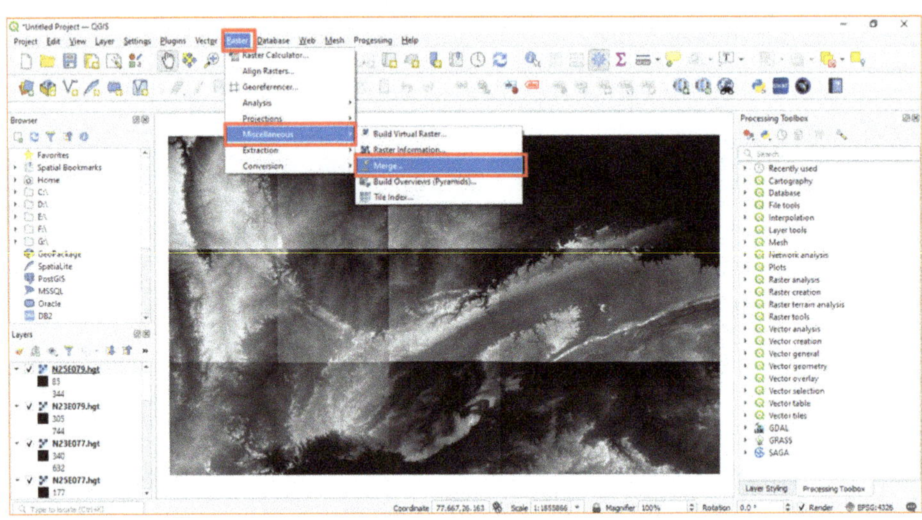

After clicking on the **Merge** option, this dialog box will appear. Click on the **Input Layers** option.

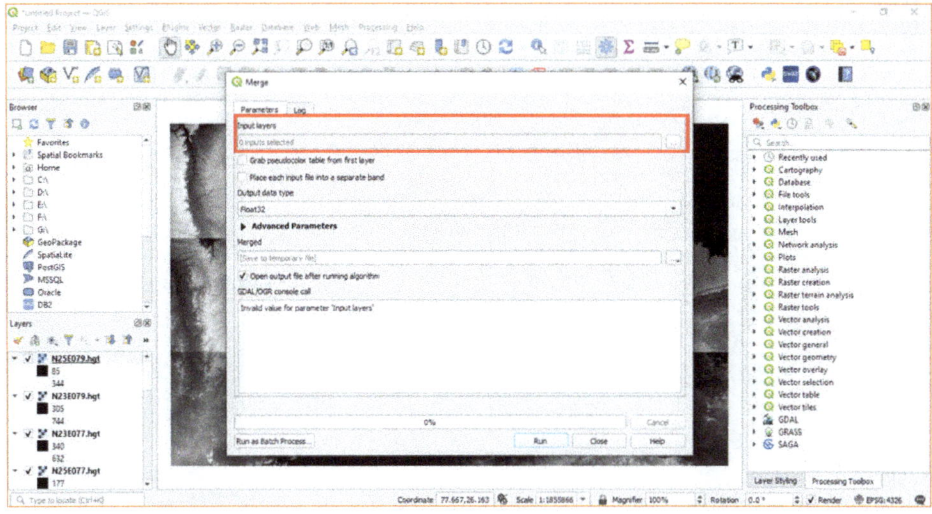

Choose the **Select All** option. Ensure that all downloaded tiles are checked. Then click on **OK** option.

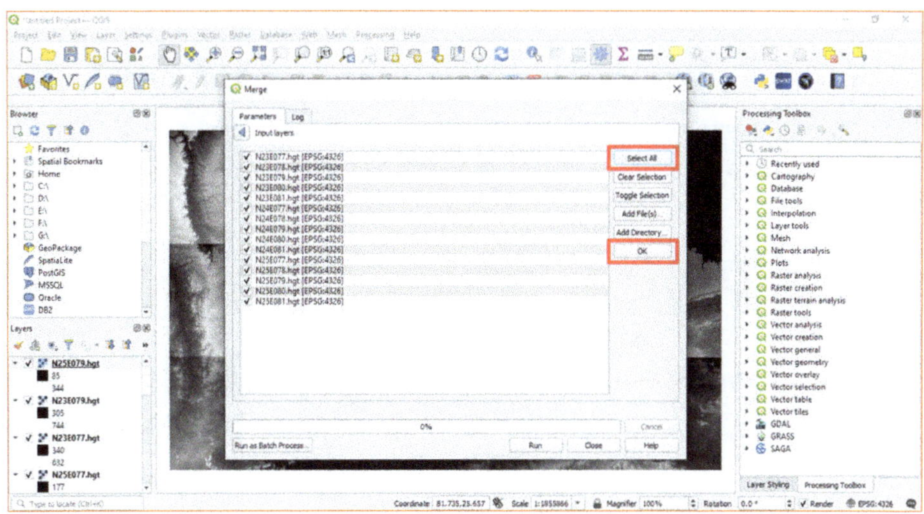

Go to the **Output data type** option and select the **Float32** option from the drop down list. The output data type may be different according to applicability.

Then go to **Merged** option and click on **Save to File** option.

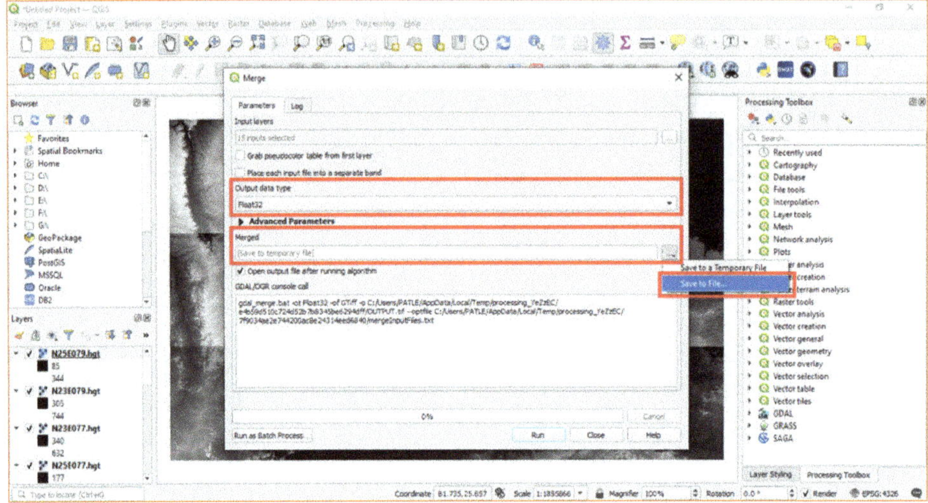

Browse the folder, where you want to save the merged DEM file. Give the file name like **DEM.tif** and **save** it.

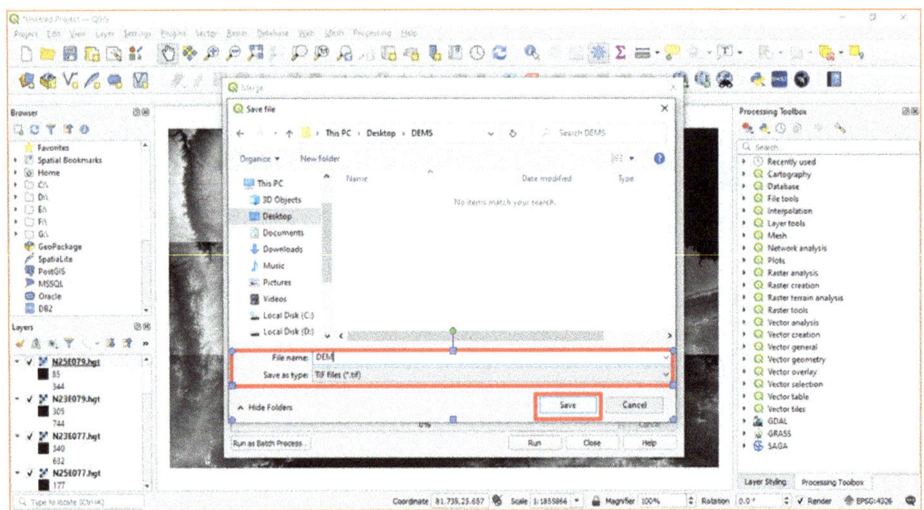

Click on the **Run** option.

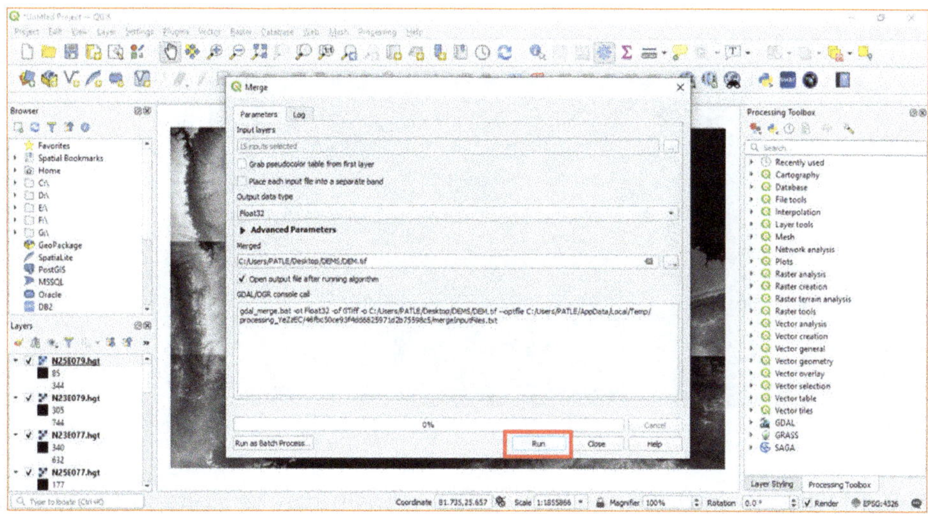

After completing the processing algorithm is 100%, then click on the **Close** option.

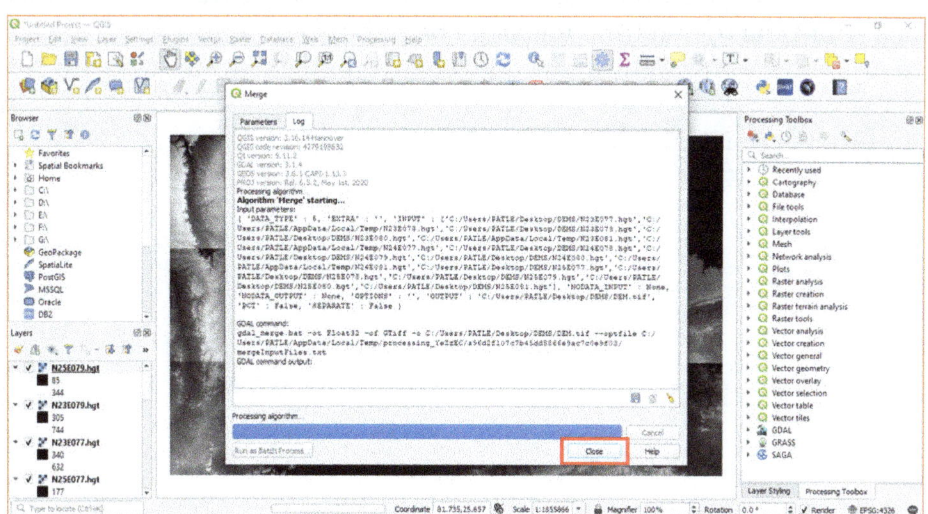

DEM has been merged in a single file. The maximum and minimum elevation of the merged DEM is display in **Layer** panel.

Note: Delineation of Watershed or River Basin can be done in QGIS and ArcGIS both the software. In this study, ArcGIS ® 10.8 software has been used to delineate the watershed/ river basin for the study.

Firstly, Open the **ArcMap 10.8** software and click on the **Add Data** icon and then **Add Data** option from the menu bar.

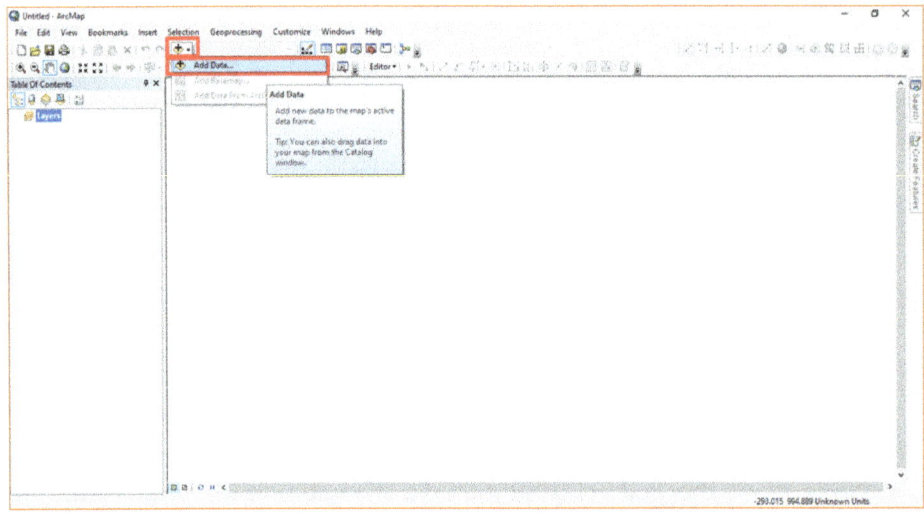

Browse the folder containing DEM data. Select the **DEM.tif** and click on **Add** option.

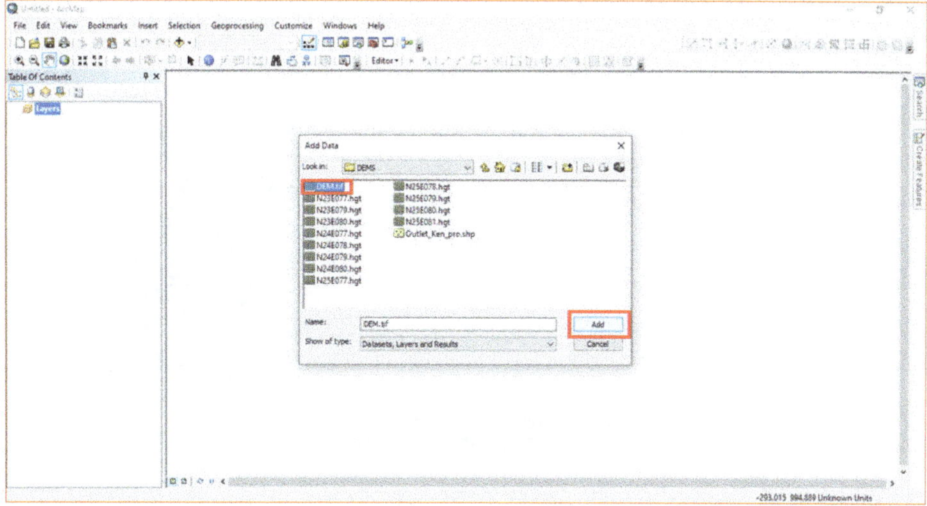

DEM.tif data has been added and it is showing in **Layer** panel. Click on **Search** menu and type **Project Raster** then click on **search** icon. After clicking, results will be appeared. Double click on **Project Raster (Data Management)** option.

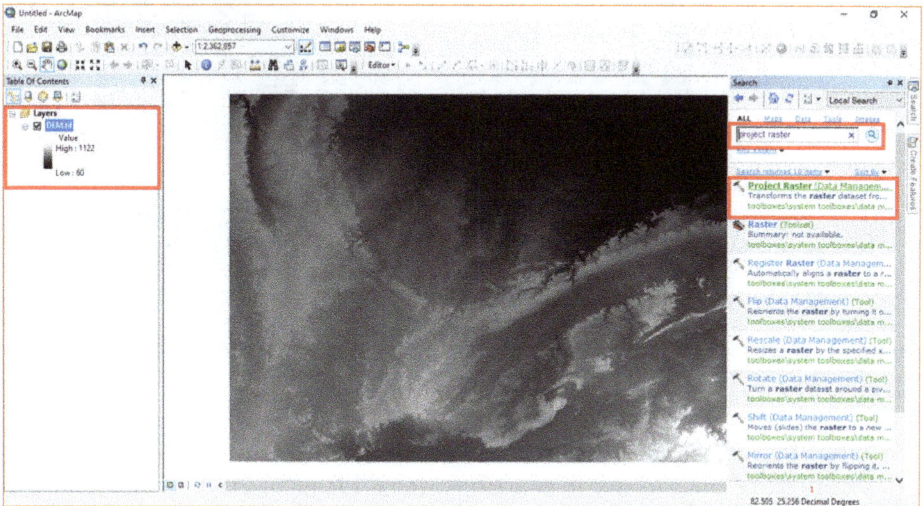

After that, Select the **Input Raster** from the dropdown menu. Give the output raster name **DEM_Pro** with extension **.tif** in **Output Raster Dataset** option. The Output Coordinate System is selected **WGS_1984_UTM_Zone_44N** as projected coordinate system. It may be different according to study area.

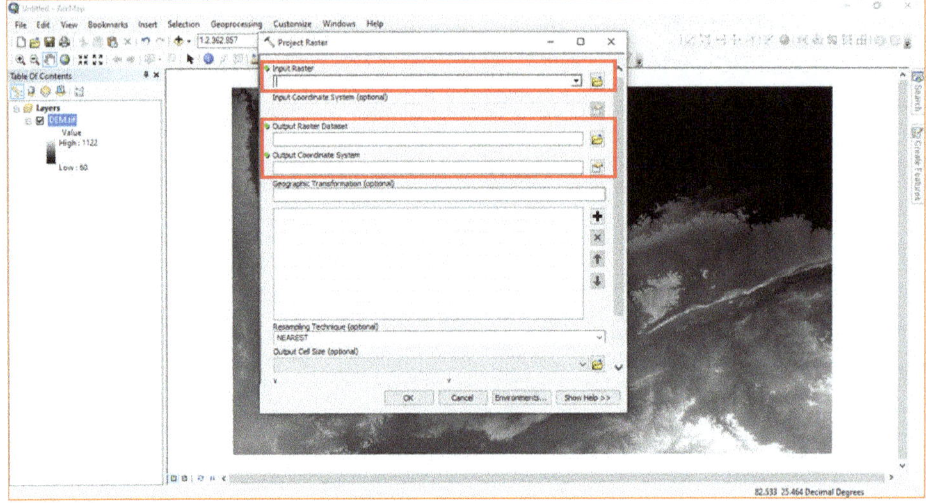

Click on **OK** option.

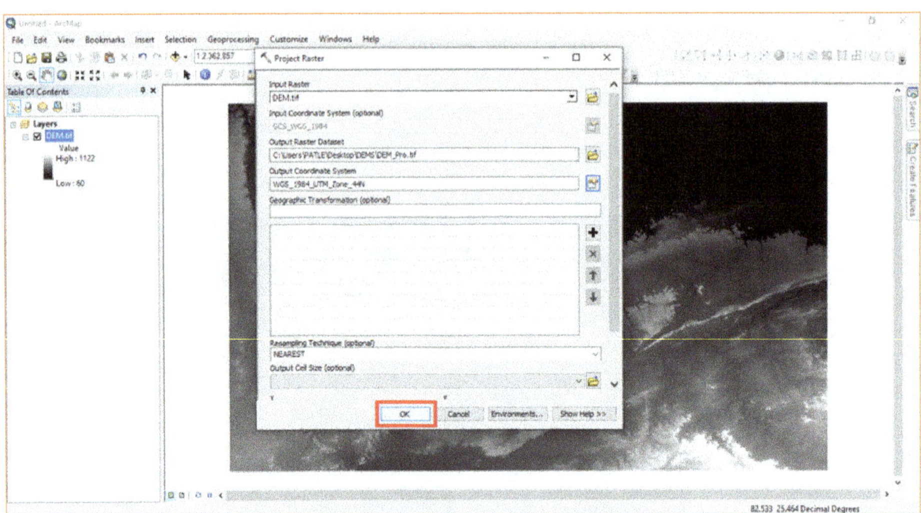

The DEM has been projected. It looks little bit tilted. In the right bottom, lat-long values are displaying in metric system/ Projected Coordinate System.

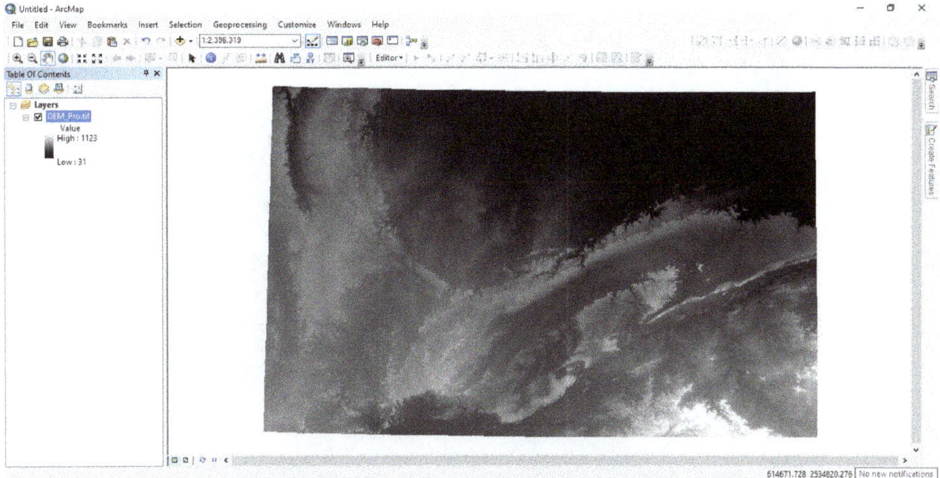

Click on the **ArcTool Box** icon.

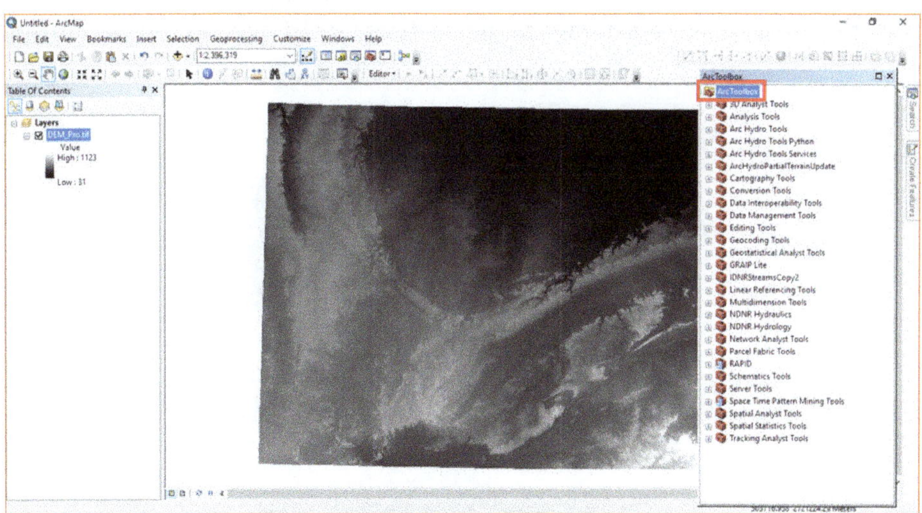

In the ArcTool Box, go to **Spatial Analyst Tool** > **Hydrology** > **Fill**. Double click on **Fill** option.

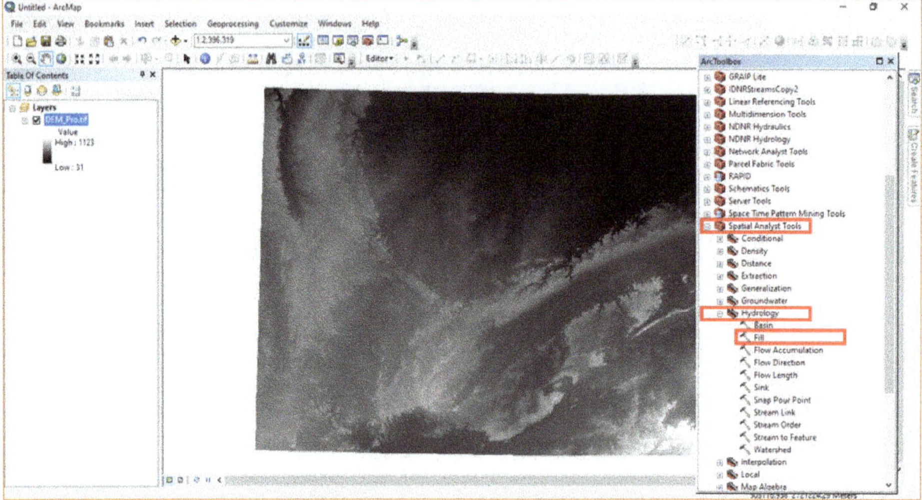

Click on the **Input surface raster** option and select the **DEM_pro.tif** file. Give the **Output surface raster** name such as **Fill.tif** and hit the **OK** button.

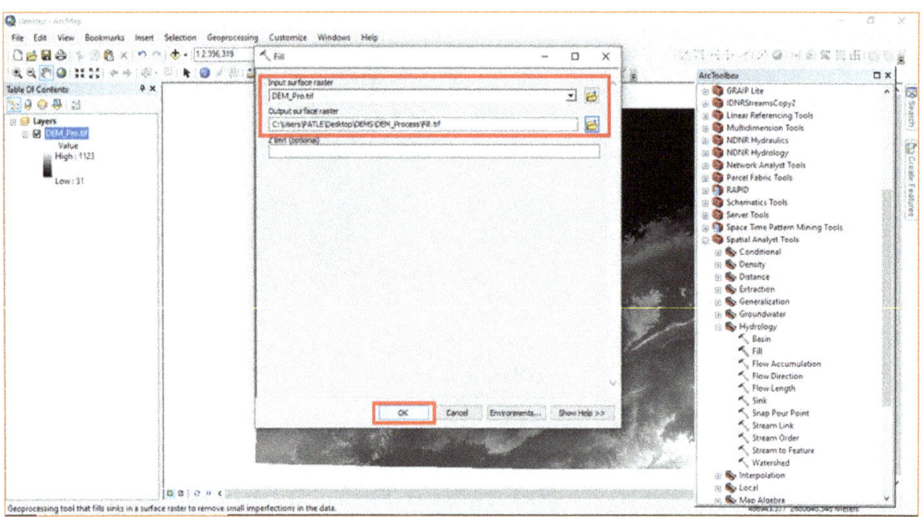

Fill dem has been generated. Go to **Spatial Analyst Tool** > **Hydrology** > **Flow Direction**. Double click on the **Flow Direction** option.

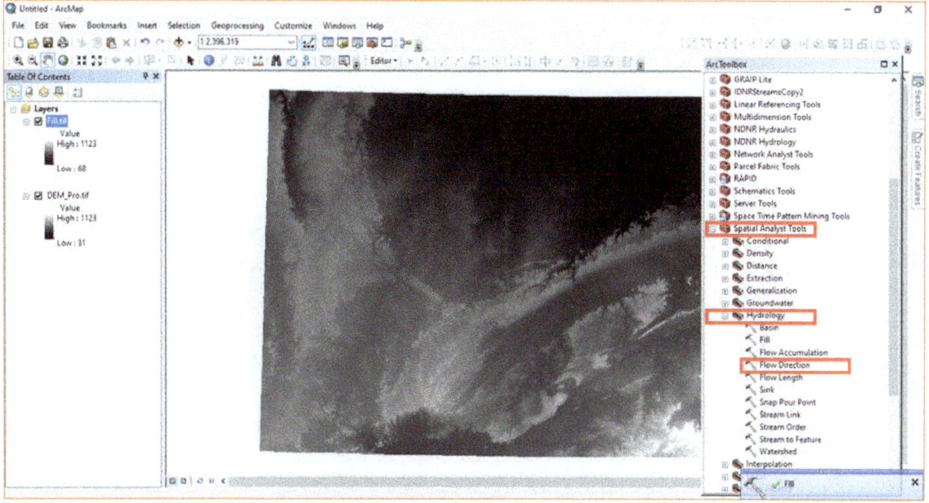

Select the **Fill.tif** file as an input raster surface. Give the output flow direction raster name as **Flow_Dir.tif**.

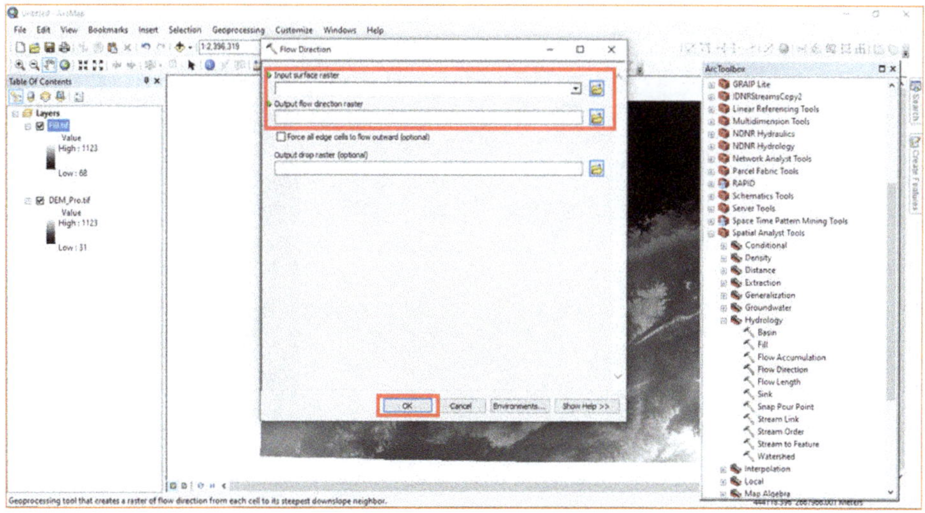

Flow Direction has been generated. Pixel value 1 to 128 depicts the different direction of flow according to ESRI code.

Go to **Spatial Analyst Tool** > **Hydrology** > **Flow Accumulation**. Double click on the **Flow Accumulation** option.

Select the **Flow_Dir.tif** file as an Input flow direction raster. Give the output accumulation raster name as **Flow_Accu.tif**. Click on **OK** option.

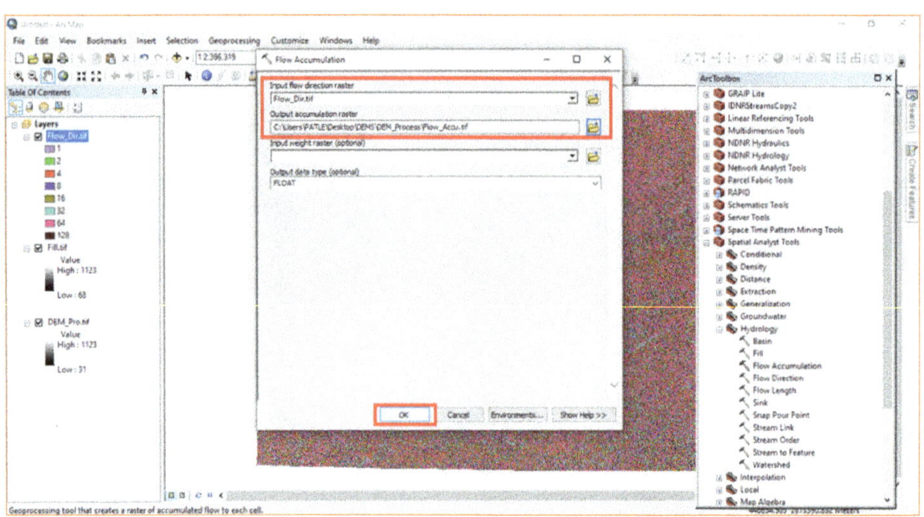

Flow accumulation file has been produced.

Go to **Spatial Analyst Tool** > **Map Algebra** > **Raster Calculator**. Double click on the **Raster Calculator** option.

Here, it is need to provide the condition, so that hanging streams will remove and all the streams could be generate having greater than 250 m lengths.

Double click on **Con** option from the **conditional** section. Prepare the statement given as below:

Con("Flow_Accu.tif" >= 250,1,0)

Then, give the output raster name as **condition.tif** and click on **OK** option.

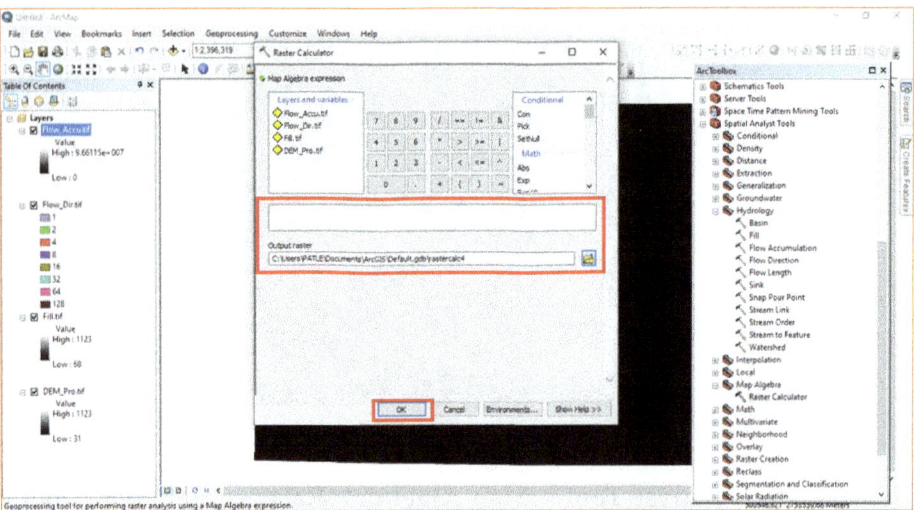

Condition raster has been generated. In this raster pixel value1 represent the streams.

Create the point shape file of outlet of the river basin. Then, Add the outlet shape file in ArcMap 10.8.

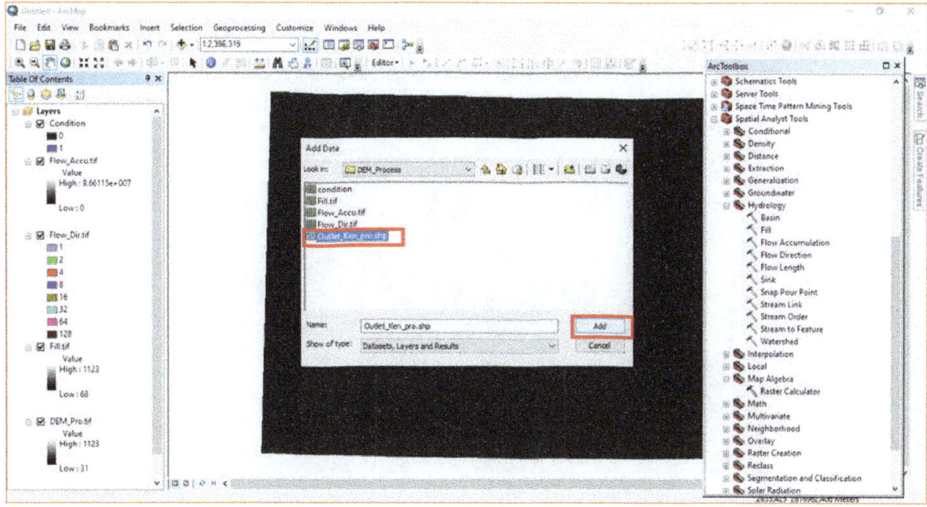

Outlet of Ken has been added.

Zoom-in the outlet point of the river basin. We can see that Outlet is little away from the stream showing in the condition raster. Outlet point should be just over the any of the pixel of stream. It needs to create again outlet point just upon the stream pixel.

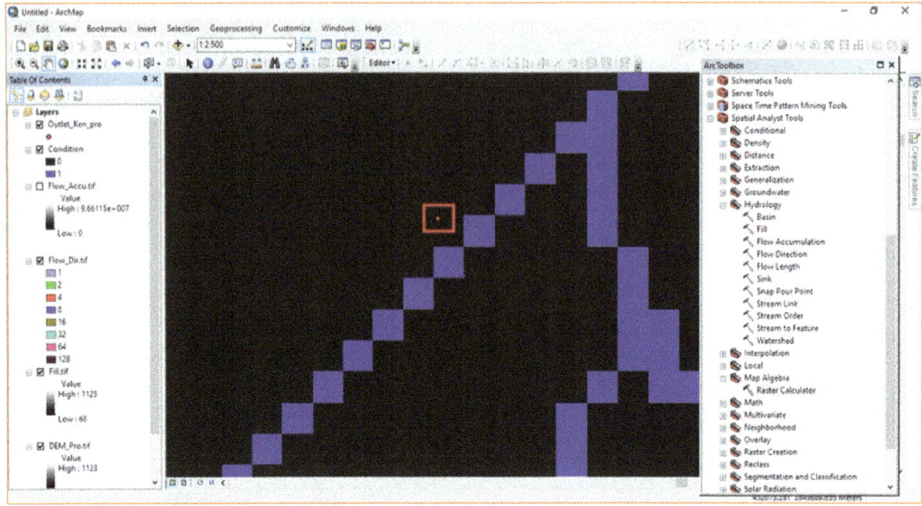

Click on the **Catalog** icon and go to **Folder Connections** and select the folder where you want to save the outlet point shapefile. Right click on the folder, click on **New > Shapefile**.

Give the shapefile name is **Pour_Point.shp** and feature type is **Point**. In Spatial Reference section, click on **Edit** option. Add the Geographic Cordinate System: **GCS_WGS_1984**. Then provide the Projected Coordinate System: **WGS_1984_UTM_Zone_44N.** The projected coordinate system may be different according to location of the study area. Finally, Click on **OK** option.

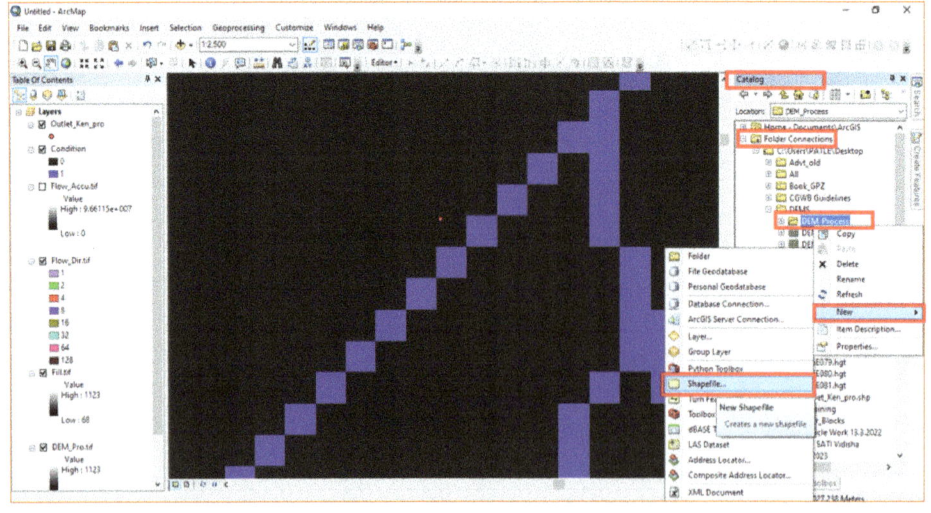

Click on the **Editor** option and select the **Start Editing**.

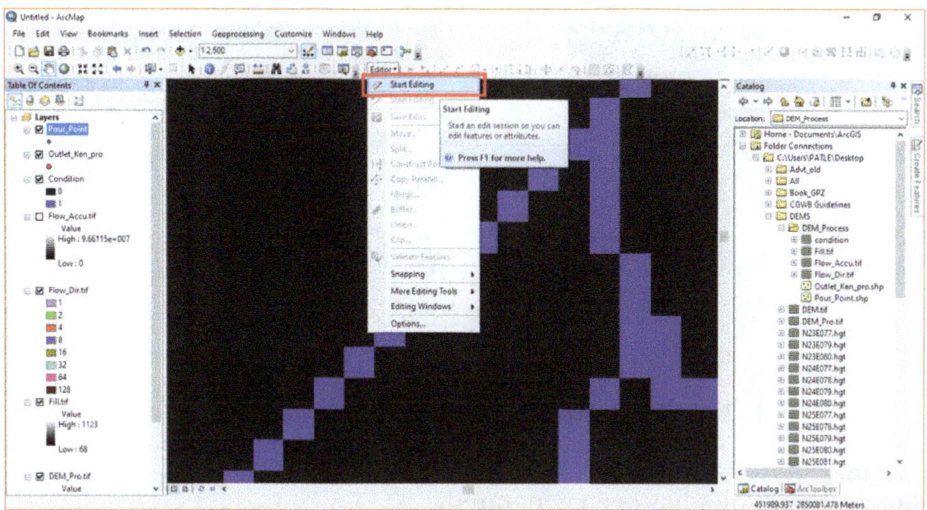

After Start Editing, select the **Pour_Point** and Click on **OK** option. Then, click on **Create Feature** option and select the **Point** option.

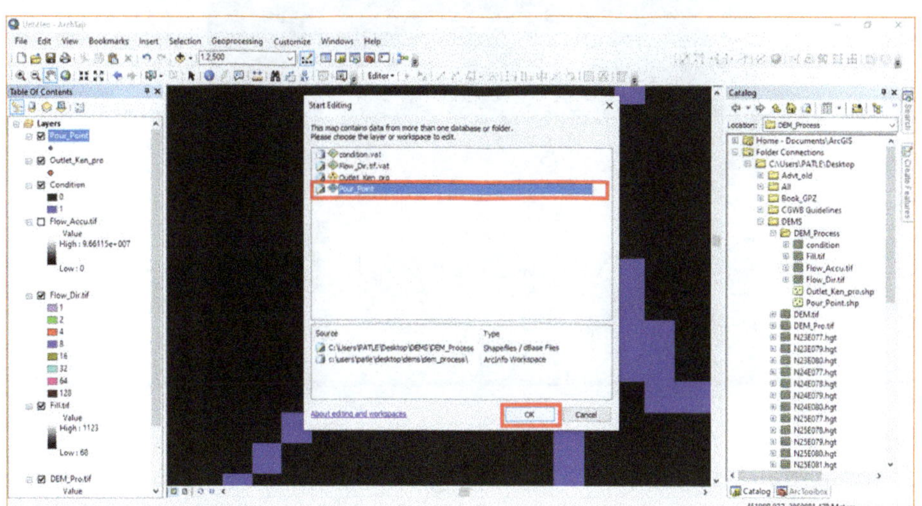

Just click upon the pixel near by the outlet and click on the **Save Edits** then **Stop Editing**.

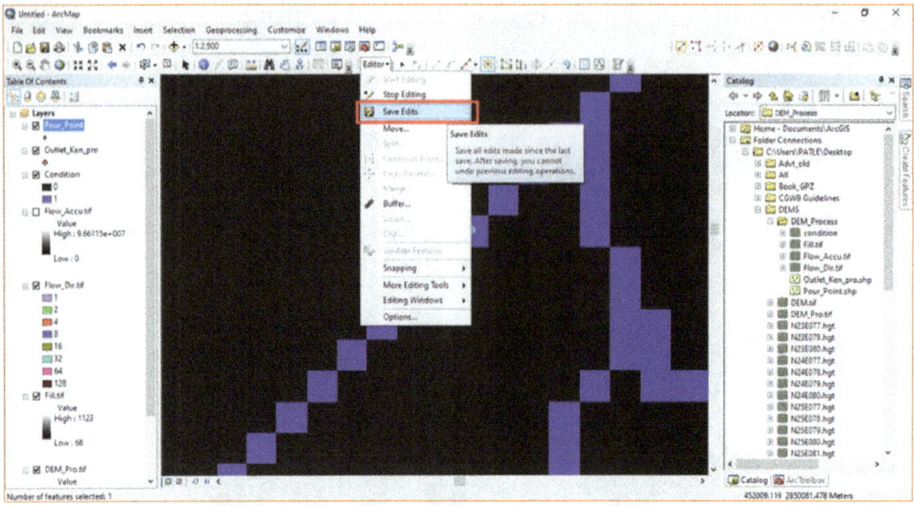

Finally, Pour Point has been generated.

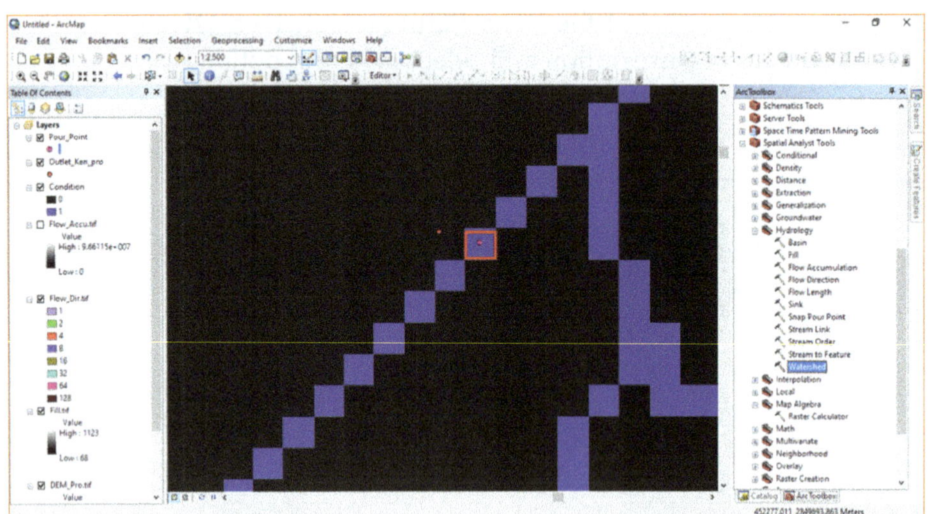

Go to **Spatial Analyst Tool** > **Hydrology** > **Watershed**. Double click on the **Watershed** option.

Select **Flow_Dir.tif** as an Input flow direction raster, **Pour_Point** as an Input raster or feature pour point, and give the output raster name as **Watershed**. Click on **OK** option.

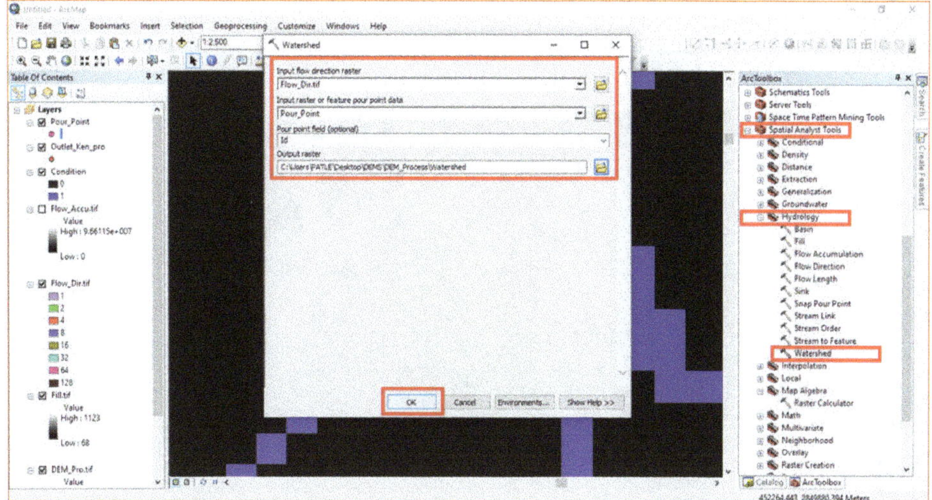

Watershed raster has been produced.

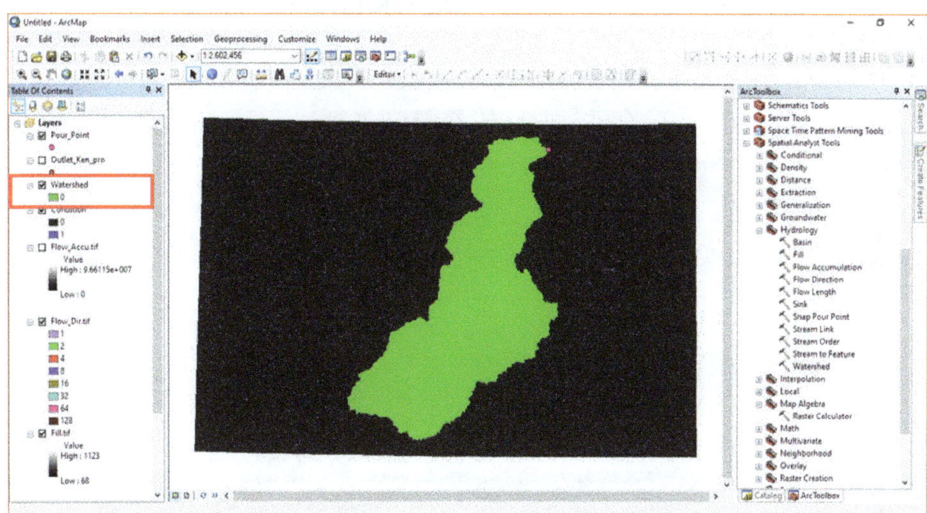

Go to **Conversion Tools > From Raster > Raster to Polygon**. Double click on the **Raster to Polygon** option.

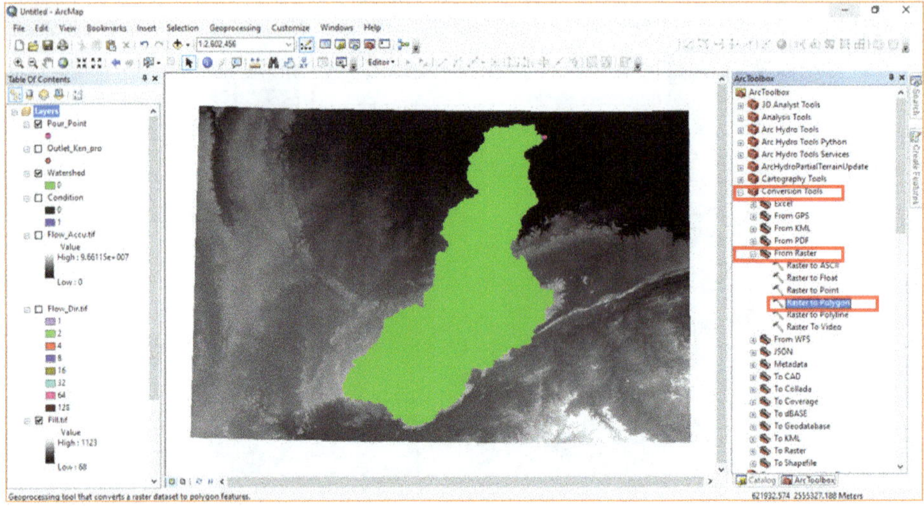

Watershed file selected as Input raster. The output polygon features name given as **Ken_Basin.shp** and presses the **OK** button.

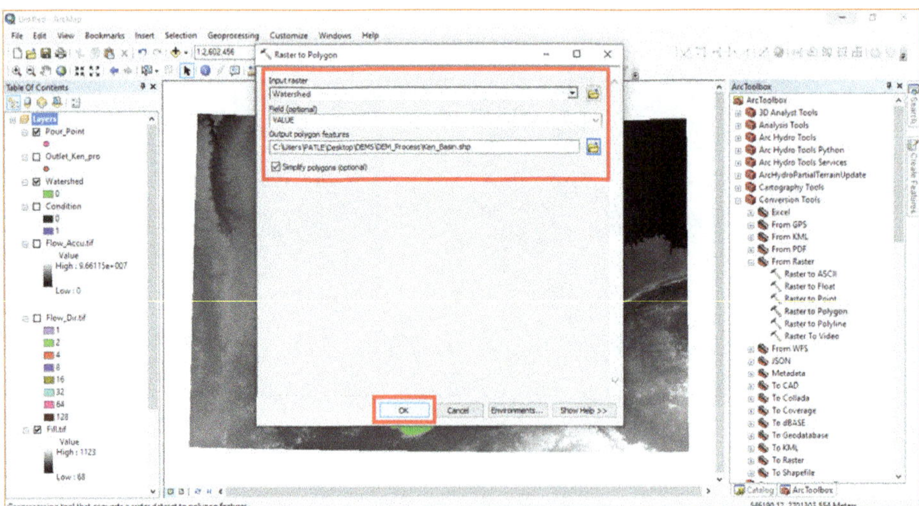

Base map of Ken River Basin or Basin Boundary has been developed.

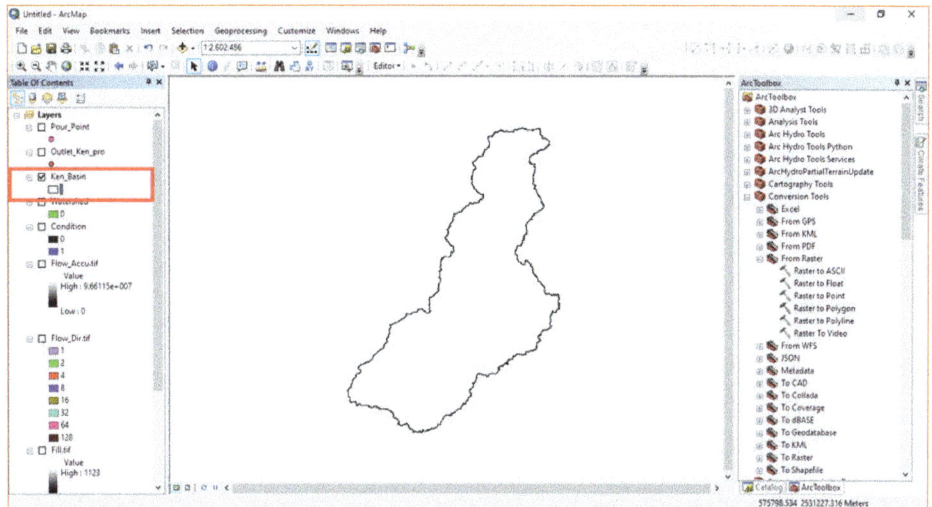

AREA CALCULATION

Right click on the file name of **Ken Basin** and click on **Open Attribute Table** option

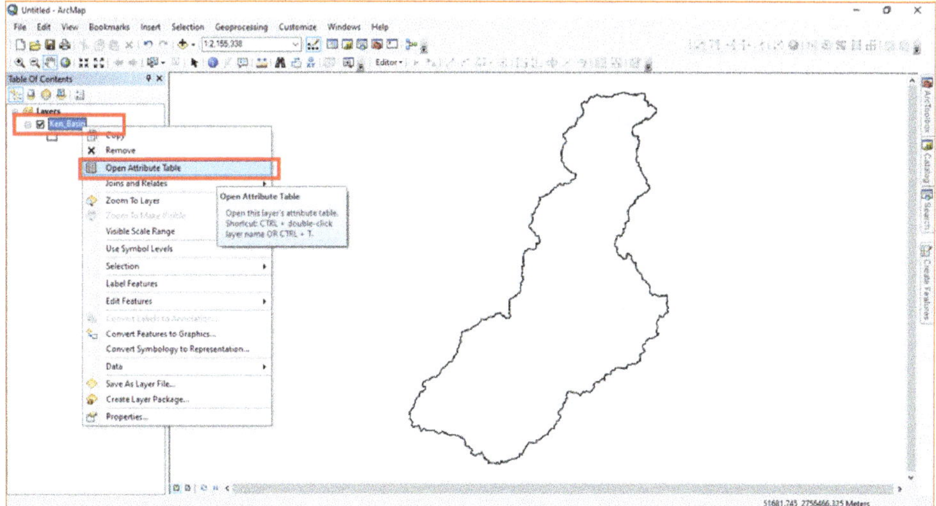

Click on the **Table Option** which is available in Top left corner of the Table.

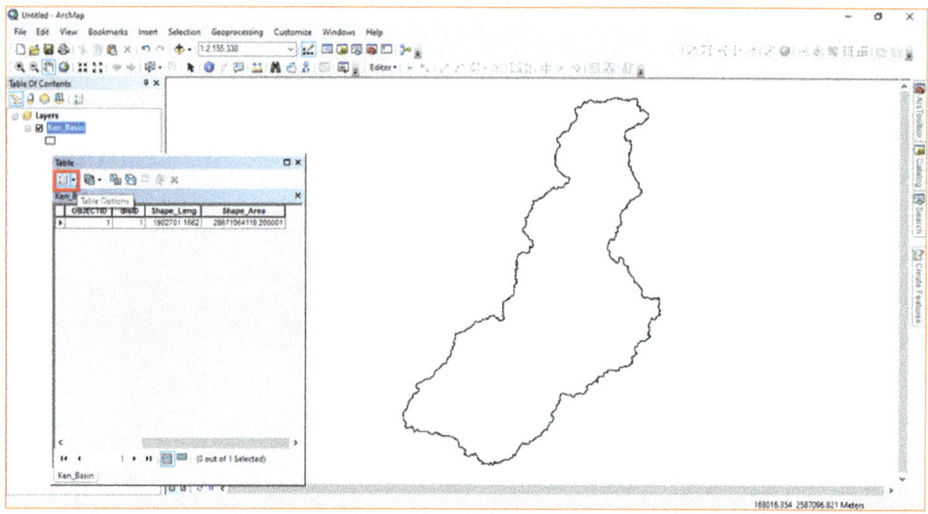

Click on the **Add Field** option.

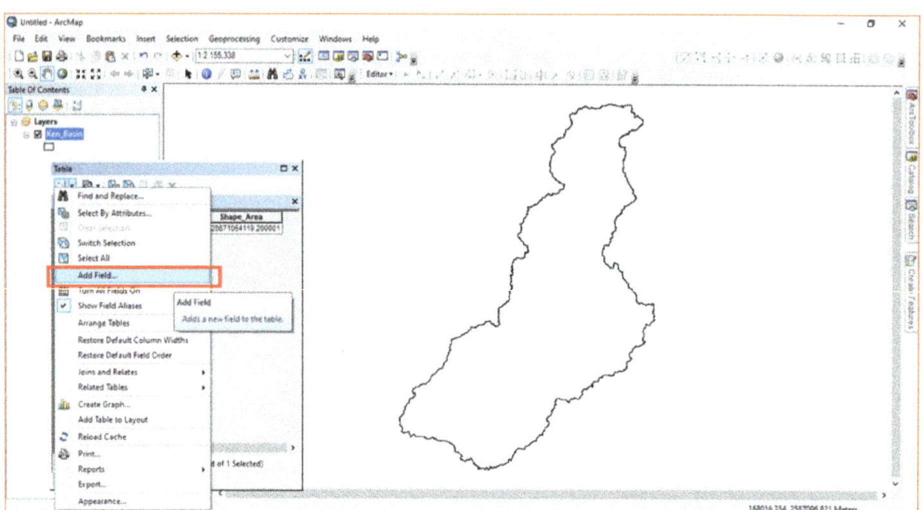

Give the Name: **Area_km2** and Type: **Float**. Type may be different accordingly. Click on **OK** option.

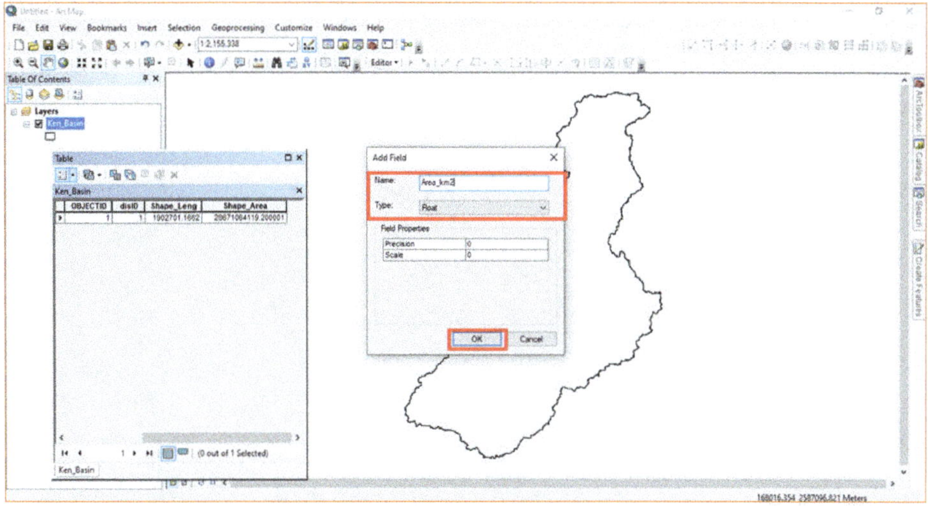

Again, **right click** on **name of Ken_Basin** and Click on **Open Attribute Table** option.

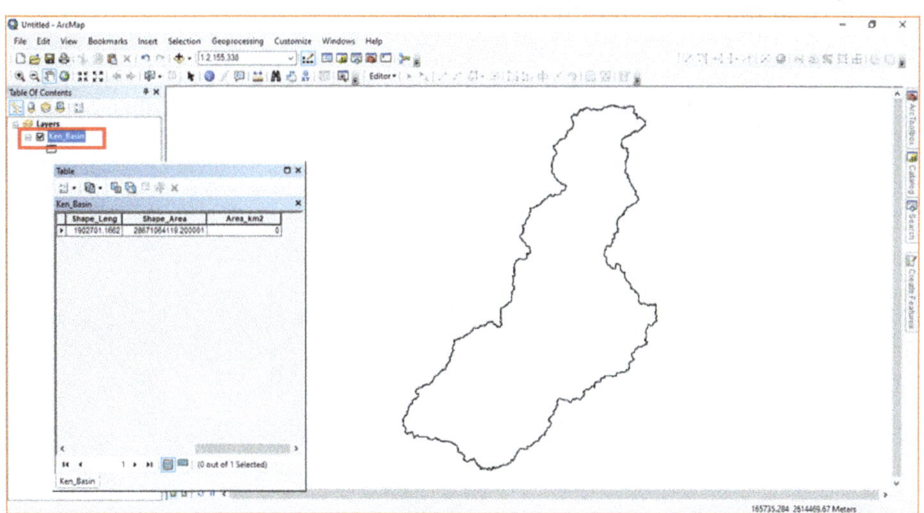

Right click on **Area_km2** alias. Then click on **Calculate Geometry** option.

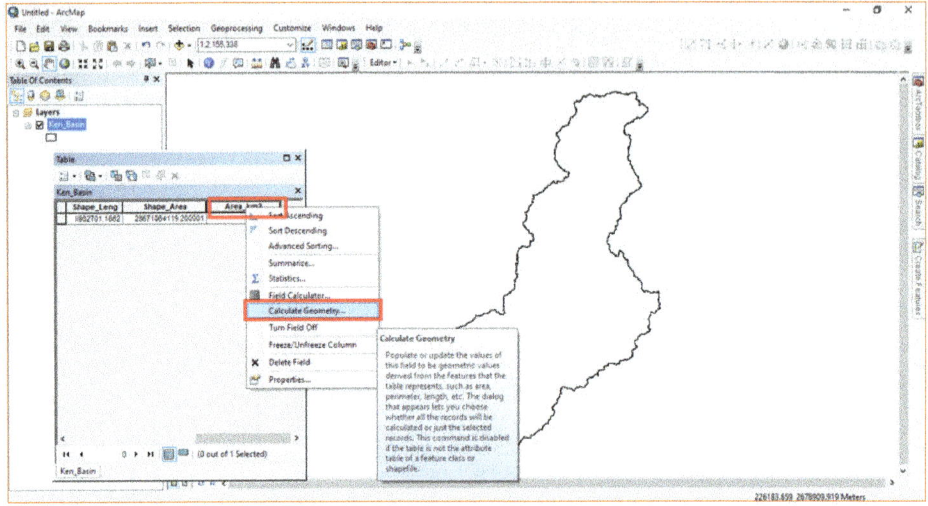

Select the property: **Area** and Units: **Square Kilometers [sq km]** then click on **OK** option.

Unit of Area may be different accordingly.

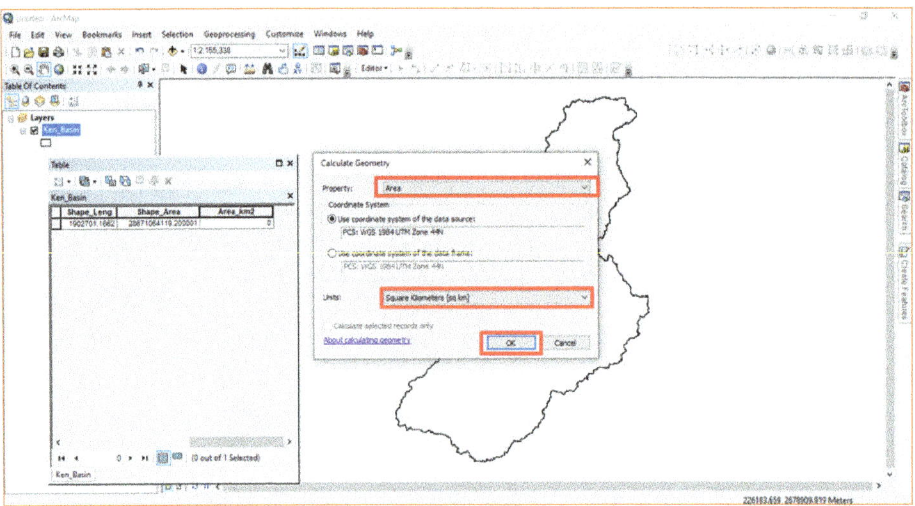

Area of the Ken Basin has been calculated.

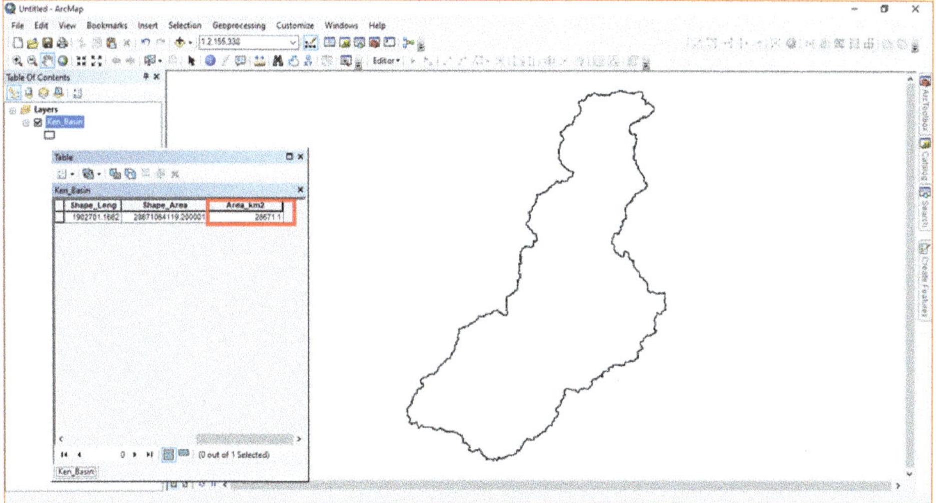

DATA COLLECTION

Groundwater influencing factors like Geology, Geomorphology, Lineaments, Land Use/ Land Cover, Soil, Rainfall, and Digital Elevation Model (for develop Slope & Drainage Density) were collected from the different geoportals.

GEOLOGY, GEOMORPHOLOGY & LINEAMENT

The geology, geomorphology and lineament data can be downloaded from the Bhukosh Portal of Geological Survey of India.

Only Registered Users of Bhukosh Portal, Geological Survey of India will be able to access Unified Download functionality in Bhukosh.

If you are a registered user (GSI Employee or External User), please login to Bhukosh (https://bhukosh.gsi.gov.in).

If you are an external user and not a registered user, please click on below link to Register:

(https://www.gsi.gov.in/webcenter/portal/OCBIS/pageQuickLinks/pageRegistration)

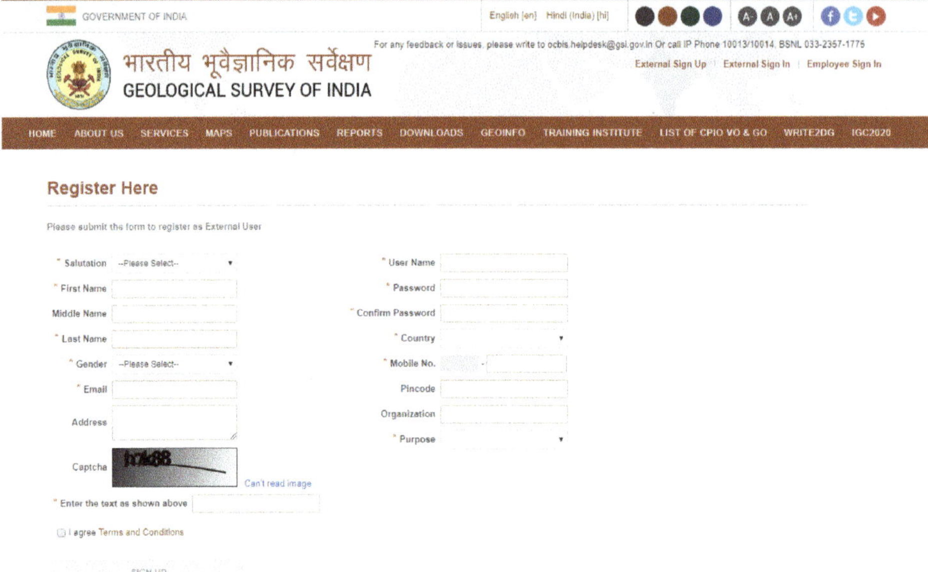

Please fill the mandatory fields and provide a valid Email-ID and Mobile Number, as it will be used for OTP verification to complete the registration process.

Once the registration is completed successfully, please login to Bhukosh (https://bhukosh.gsi.gov.in).

Fill the username and password. Click on **Login** option.

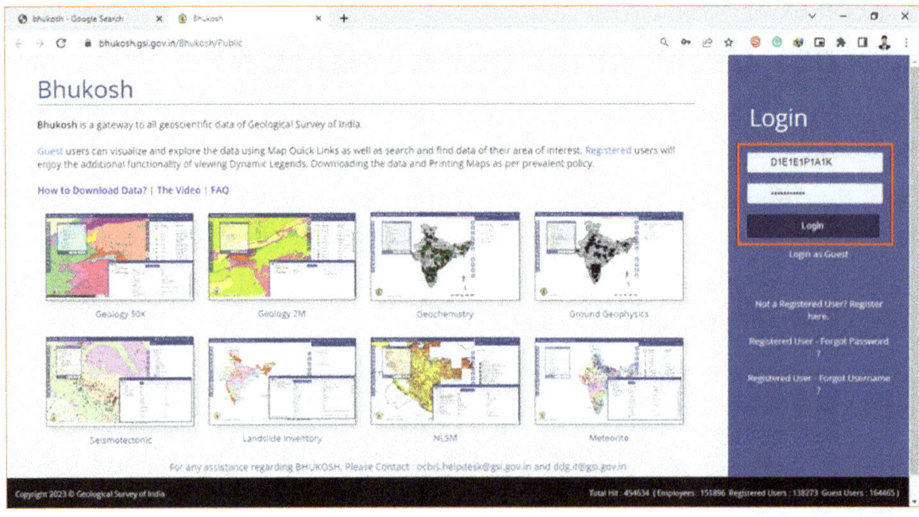

Click on **Unified Search** option and fill the area such as **States, District** or **Toposheet No**. You can use Bounding Box option according to purpose. Then, click on **Search** option to fetch the Search Results.

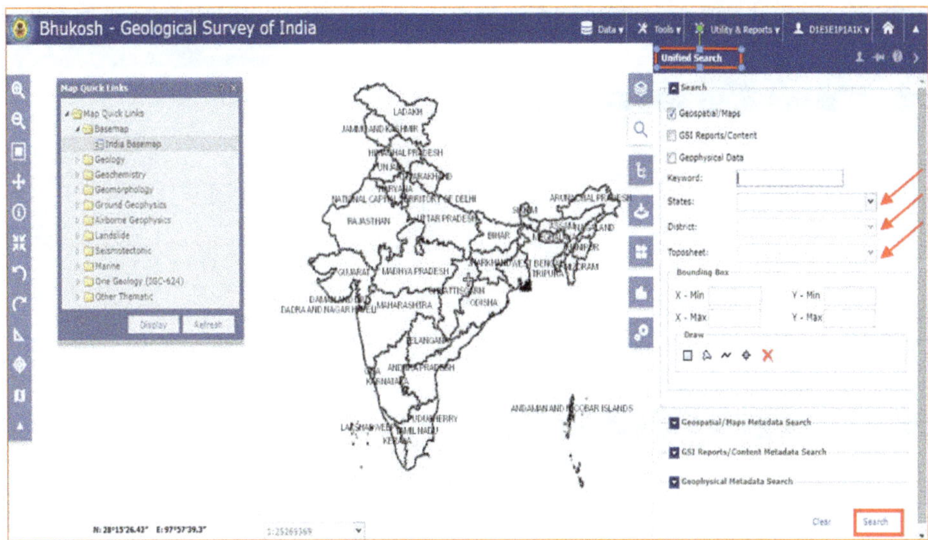

In the Search Results window, Click on **Download** icon to add the layer/ item to the Unified Download Cart. Click all Download options of the desired theme.

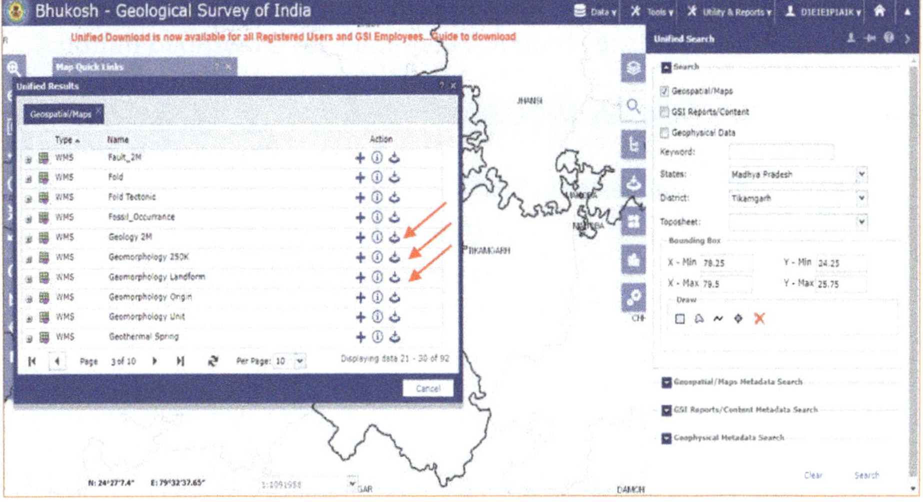

The selected Layer/Item is now added to Unified Download Cart. Similarly, more layers/items can be added to the **Unified Download Cart**. Then, click on the **validate** option.

Maximum 10 Layers/Items can be added in the Unified Download cart for each type of data – Geospatial/Maps or Geophysical Data or GSI Reports/ Content Data.

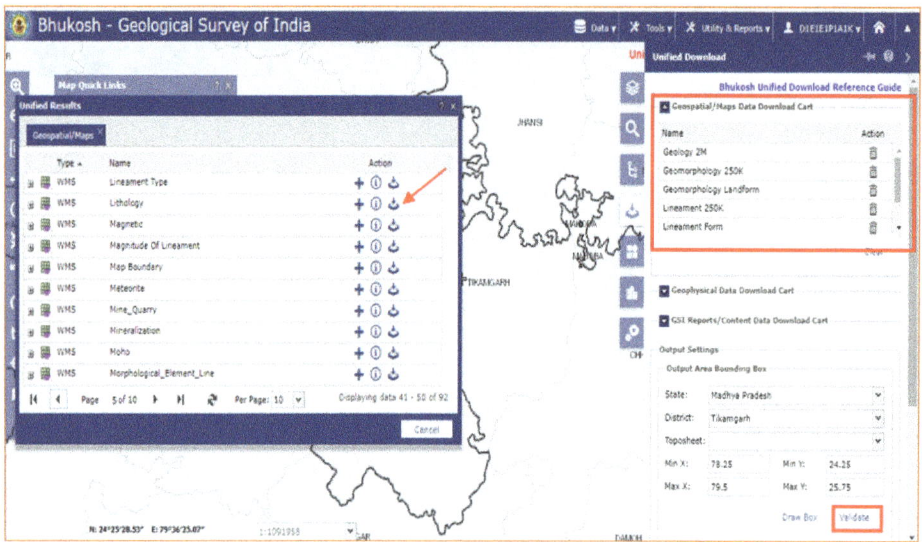

The Validate button is provided, to check if the selected layers fall inside/ outside the Output Area Bounding Box. Select the desired Output Format as either SHP (Shapefile) or GML.

Click on "**Click to see the Terms and Conditions**" to view the Terms and Conditions. Click on **Agree** button. Then, Click on **Download** button.

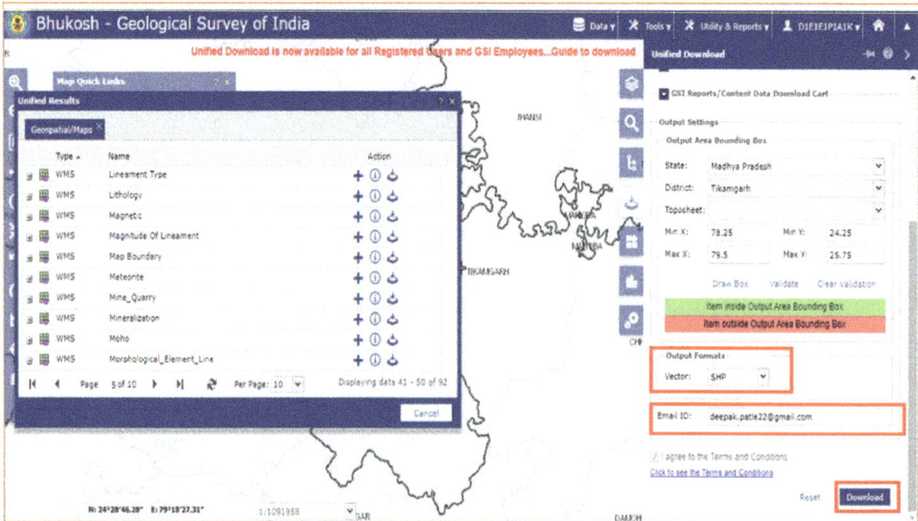

After the data download request has been processed successfully, the Download Link(s) will be to the registered **Email-ID**.

LAND USE/ LAND COVER

The Land Use/ Land Cover (LULC) map may be developed through the high-resolution satellite imagery of Sentinel-2 using supervised/ unsupervised classification method.

In this study, LULC map directly downloaded from the '**Sentinel-2 Land Cover Explorer**' developed by ESRI, Impact Observatory, and Microsoft.

Just type **esri lulc 2022 download** on google search. Click on the first results '**ESRI Land Cover – ArcGIS Living Atlas**'.

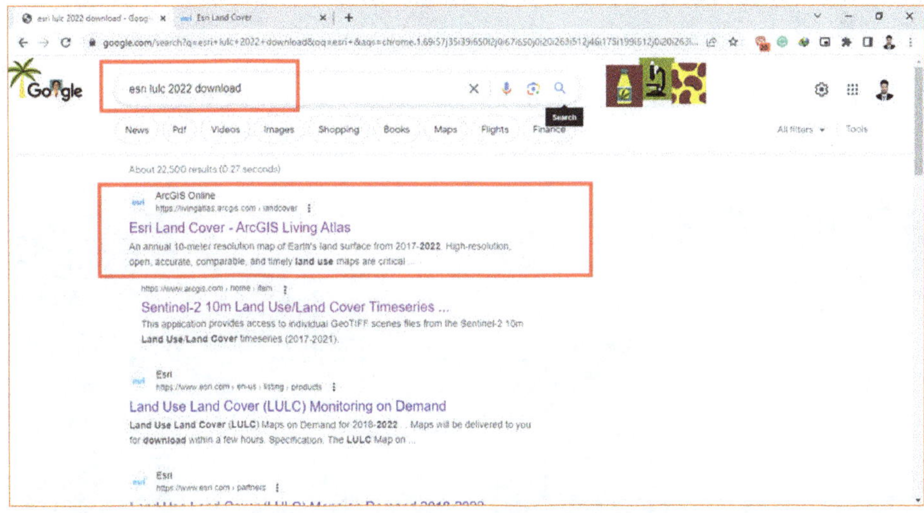

Click on the **Launch the Land Cover Explorer application**.

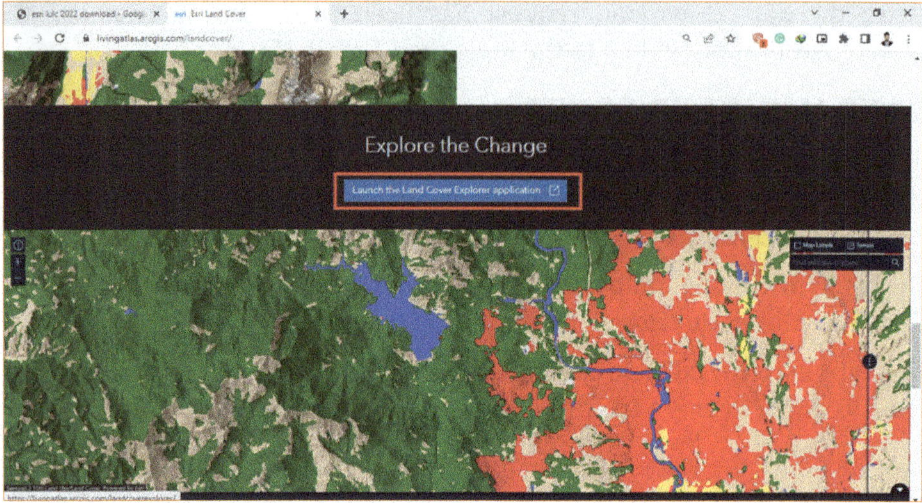

The new window will open with ESRI Sentinel-2 Land Cover Explorer.

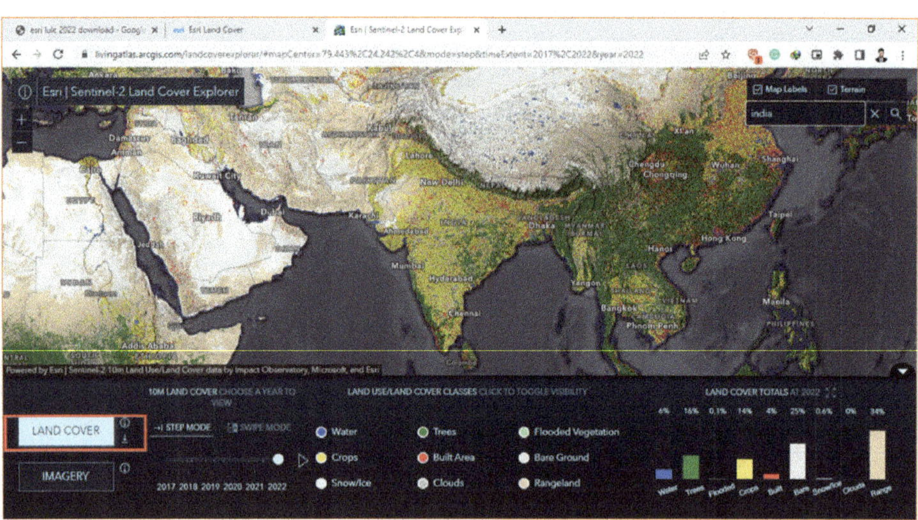

When zoom-in the map, LULC with tiles will be appeared.

You can download LULC according to year available.

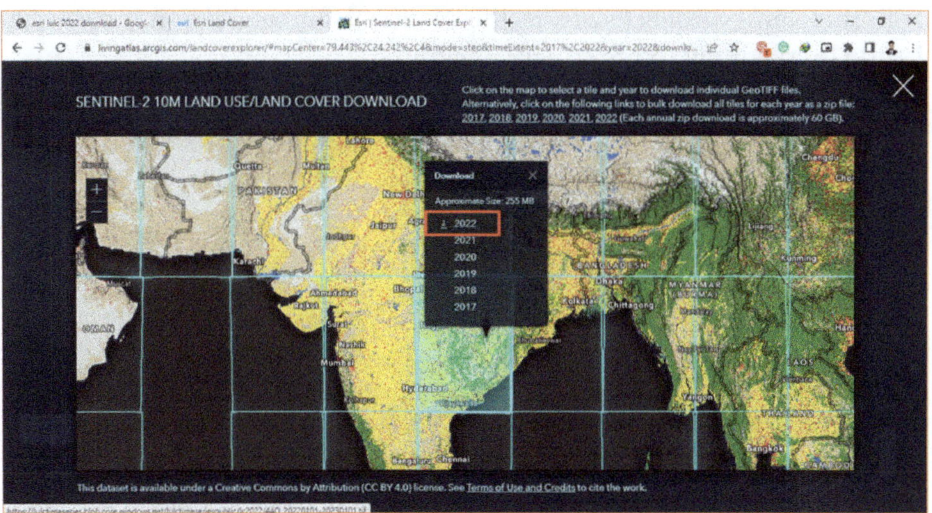

Click on the Year 2022, and LULC data will be start download.

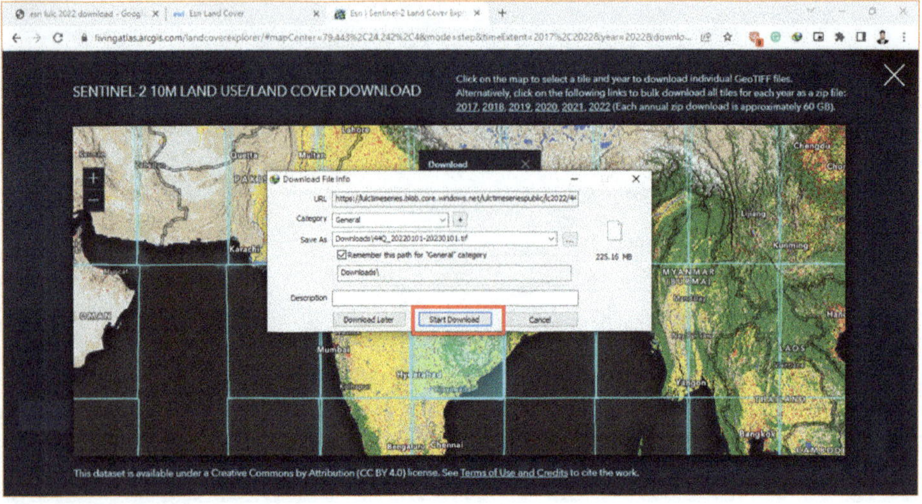

SOIL DATA

Soil data can be collected from the National Bureau of Soil Survey and Land Use planning (NBSS&LUP), Nagpur.

Soil map also available on Bhoomi geoportal of NBSS&LUP. You can extract the soil map for very small area only. But data extraction is not easy from this geoportal. (http://www.bhoomigeoportal-nbsslup.in/)

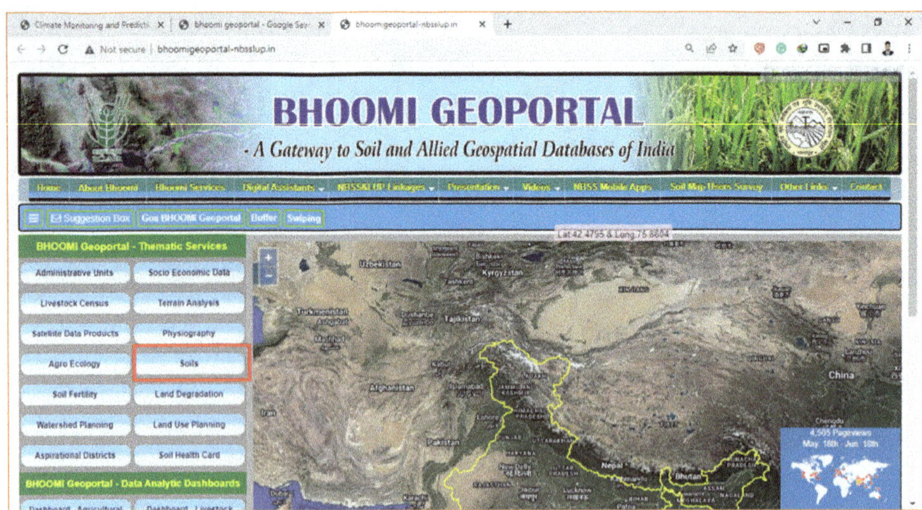

Zoom-in the map up to the study area showing. Click on any polygon/ contour. You will get the information.

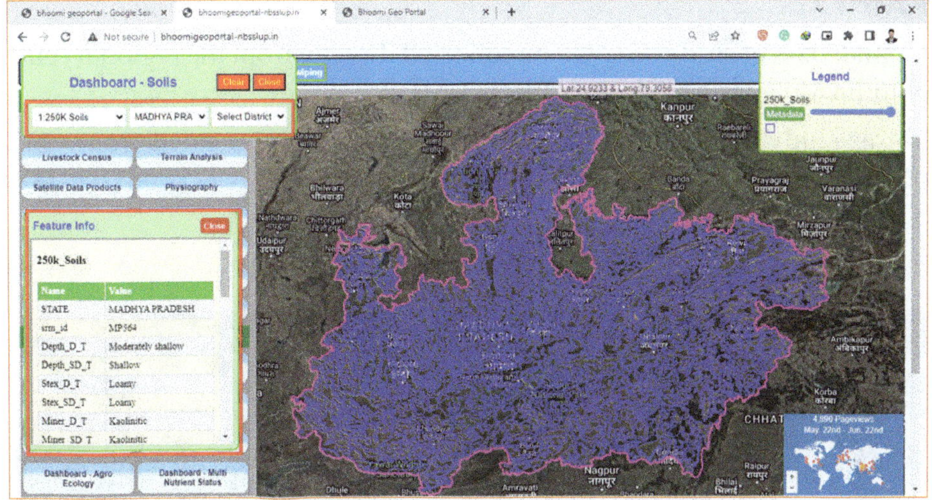

Soil data (Soil Texture) available on different scale. In this study, soil map used on a scale of 1:250,00.

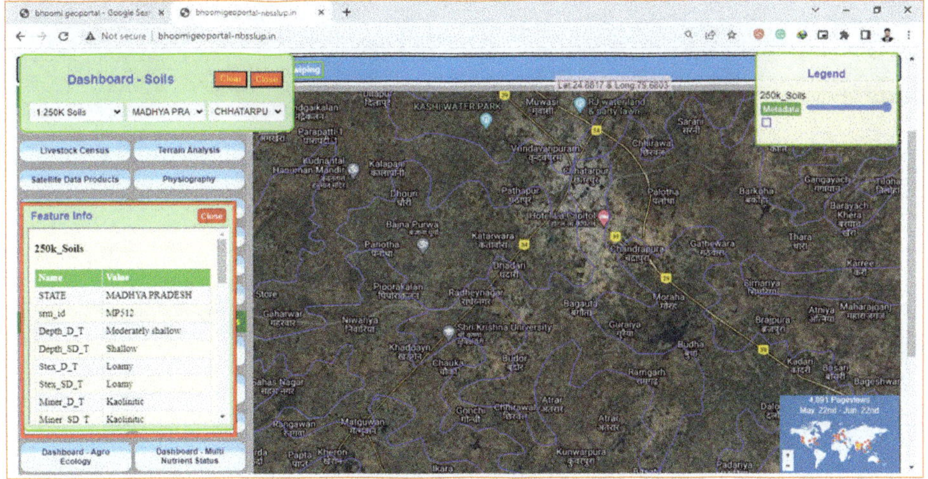

SLOPE AND DRAINAGE DENSITY

Generally, Digital Elevation Model (DEM) such as SRTM, Cartosat, and ASTER are used for the development of slope and drainage, drainage density map. SRTM 30 m DEM may be downloaded from the USGS Earth Explorer (https://earthexplorer.usgs.gov/).

In this study, different tiles of DEM were downloaded through SRTM Downloader Plugin in QGIS software. The process of DEM downloading is earlier discussed in the 'Base Map Preparation' section.

RAINFALL

Rainfall data of study area can be collected directly from the regional center of India Meteorological Department (IMD).

In this study, rainfall data for last 30 years was downloaded from CSDP portal of IMD Pune (https://cdsp.imdpune.gov.in/). Click on the **GET STARTED** option.

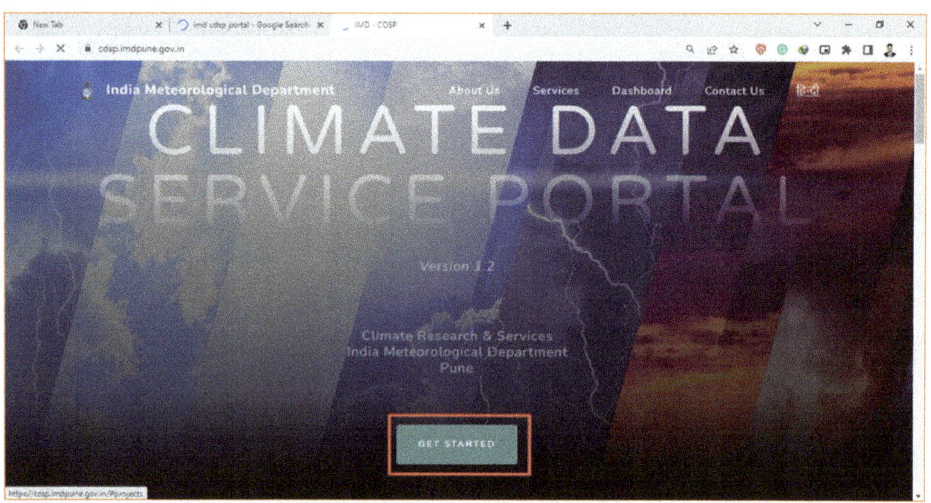

Click on the **Open Data Access** option.

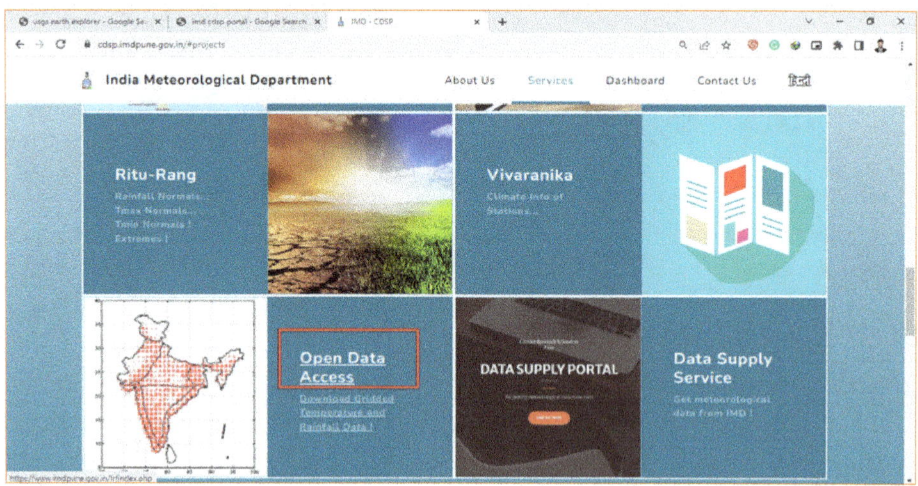

Click on the **Gridded Data Archive** from left panel and choose the **Rainfall (0.25 x 0.25) Binary** option. You can also download the netCDF file of the rainfall data.

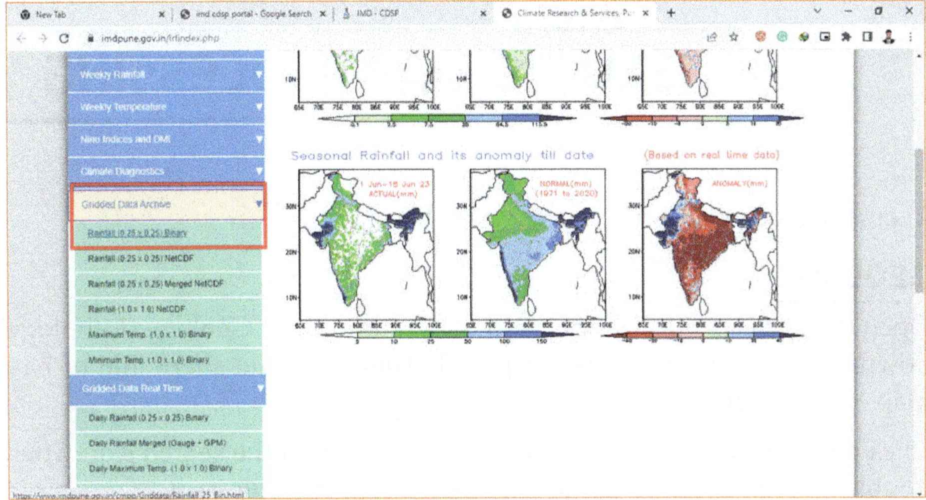

Select the **Year** and click on the **Download** option. Yearly Rainfall data will be downloaded with **.grd** file format.

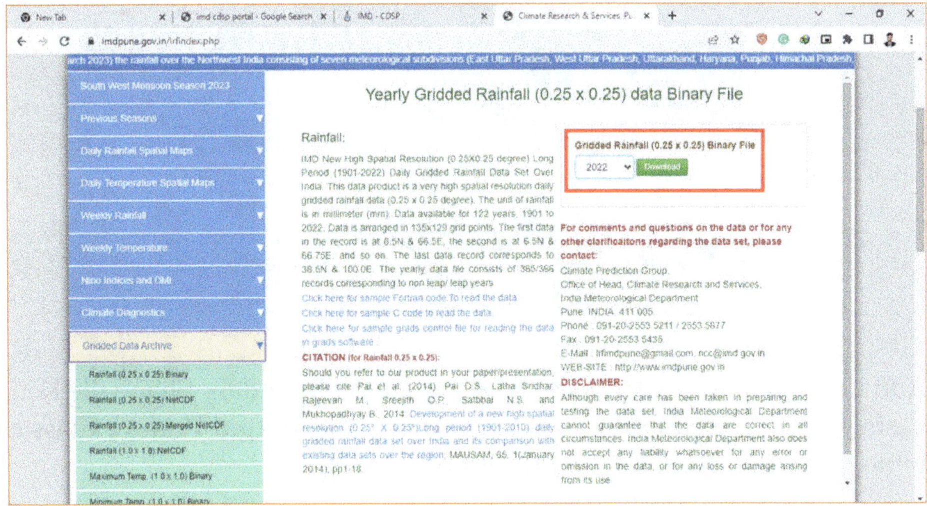

THEMATIC MAP PROCESSING

Thematic data processing is given as below:

GEOLOGY

Open the ArcMap application. Add the vector data of Geology (Lithology) downloaded from the Bhukosh portal. Then, add the shape file of base map of Ken Basin.

Click on the **Search** option and type the **clip** and hit the **search** icon. Then, double click on the **Clip (Analysis)** tool in search result item.

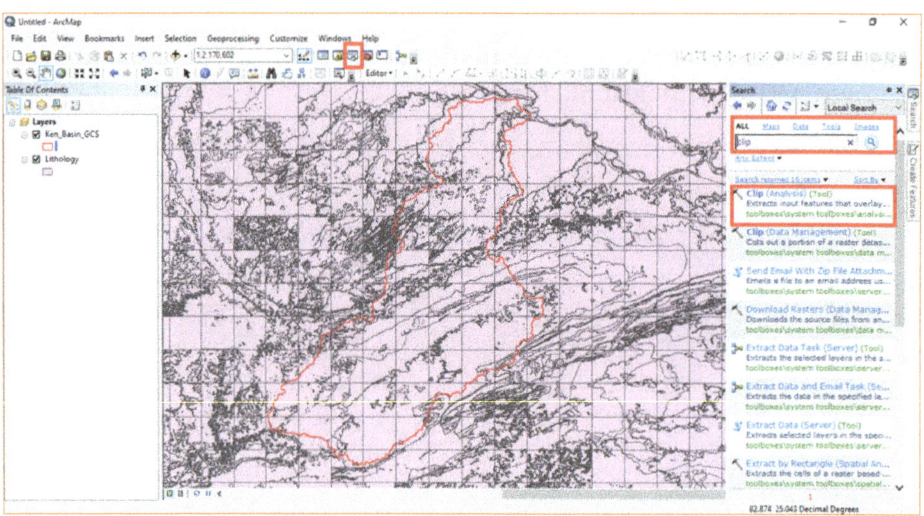

Add Geology layer as an **Input Features**, select Boundary Shape file Ken_Basin_GCS in **Clip Features** option, and give the **Output Feature Class** as Lithology_Clip. Then, hit the **OK** Button.

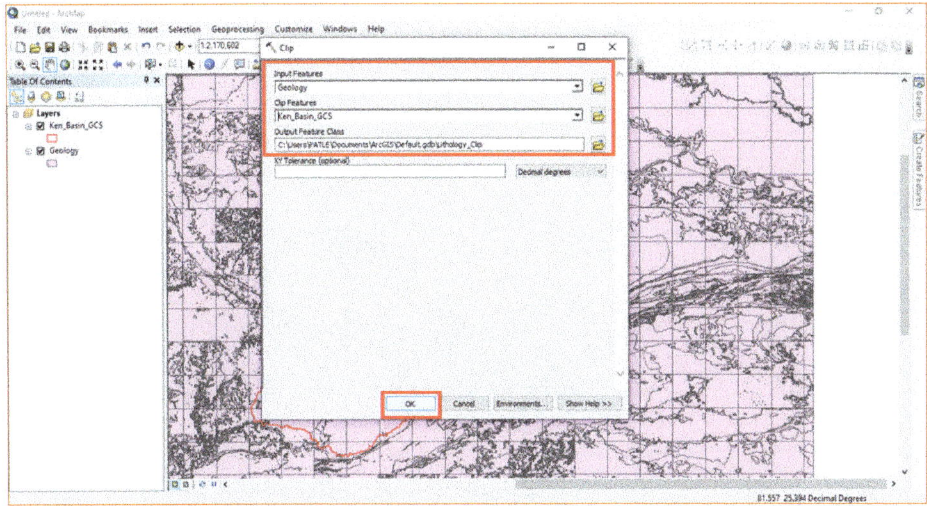

Lithology/ Geology layer has been clipped by Basin Shapefile.

In this layer, there are some rectifications are required. Polygons are merged with square polygons. It can be rectified using the reference map collected by the GSI report/ websites or any scientific literature.

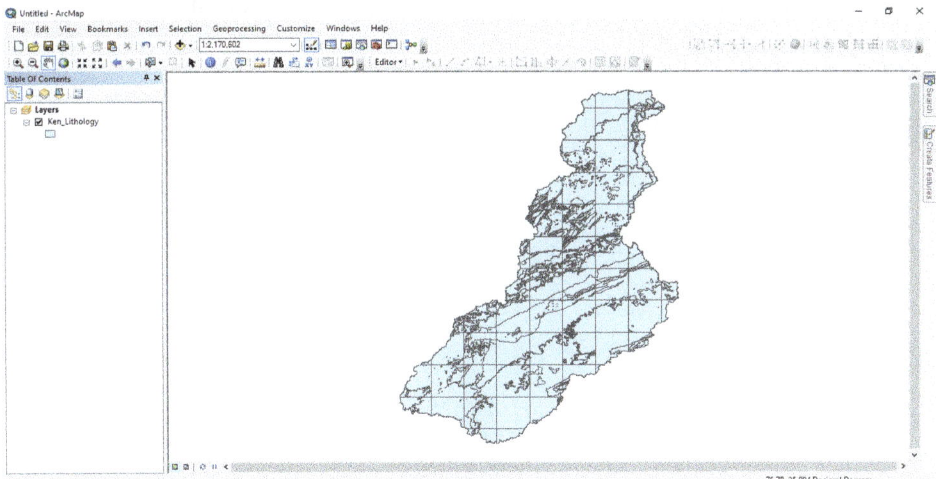

The Geology layer of Ken Basin has been rectified.

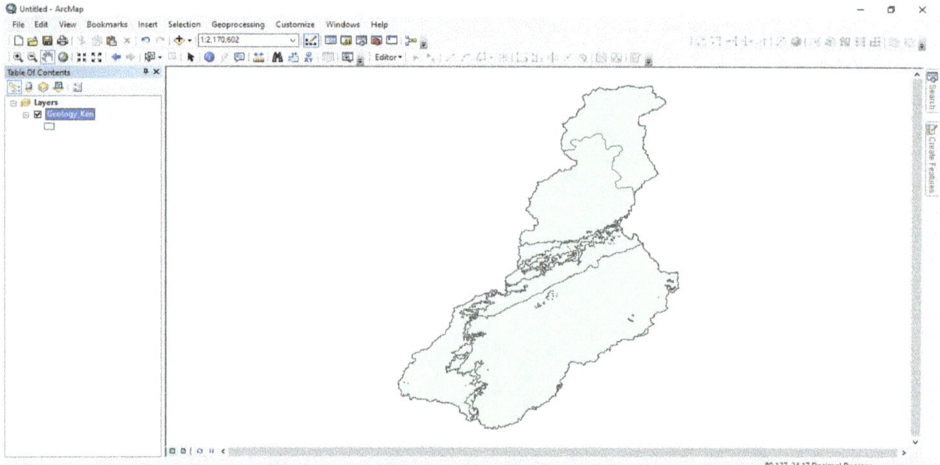

Right Click on the file name **Geology_Ken** showing in Layers panel. Then, click on **Properties** option, and go to the **Symbology** option. After this, go to:

Categories > **Value Field** > **Group Name** or **Geology** > **Add All Values** > **Apply** > **OK**

Different geology groups will be displayed in Layers panel.

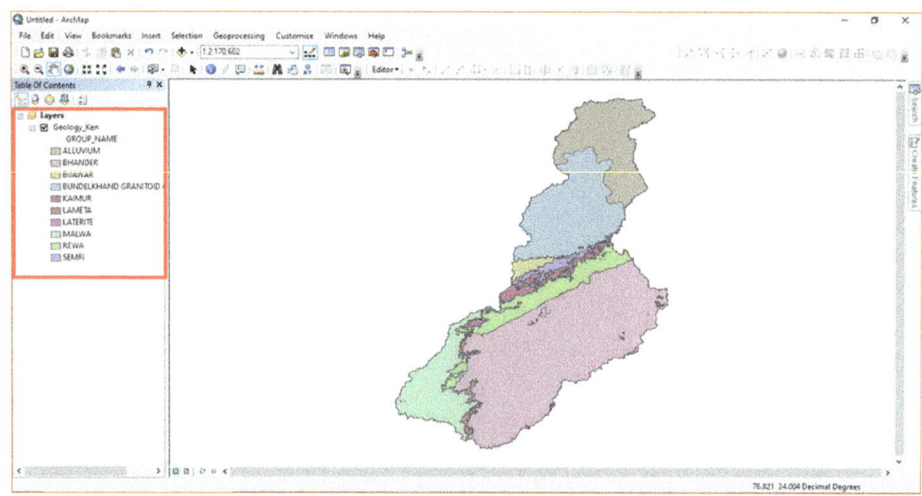

GEOMORPHOLOGY

Open the ArcMap application. Add the vector data of Geomorphology downloaded from the Bhukosh portal. Then, add the shape file of base map of Ken Basin.

Click on the **Search** option and type the **clip** and hit the **search** icon. Then, double click on the **Clip (Analysis)** tool in search result item.

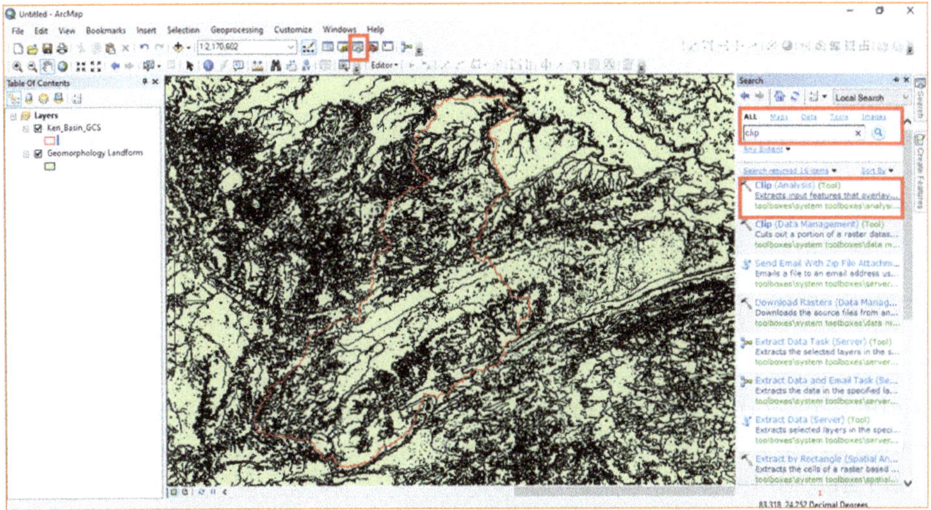

Add Geomorphology layer as an **Input Features**, select Boundary Shape file Ken_Basin_GCS in **Clip Features** option, and give the **Output Feature Class** as Geomorphology_Clip. Then, hit the **OK** Button.

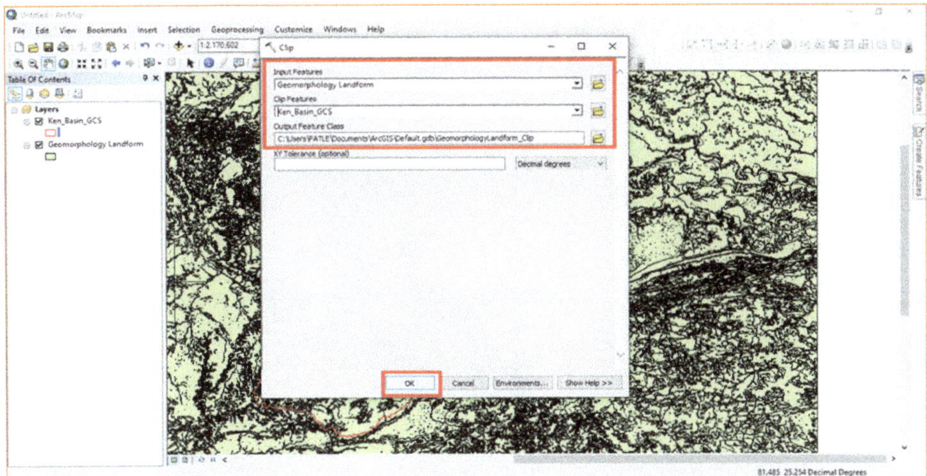

Geomorphology layer has been clipped by Basin Shapefile.

The geomorphological units are available in different classification levels such as Level-1, Level-2 & Level-3. Level-1 data are required to our study. So, the classes of geomorphology need to be regrouped according to Level-1 classification.

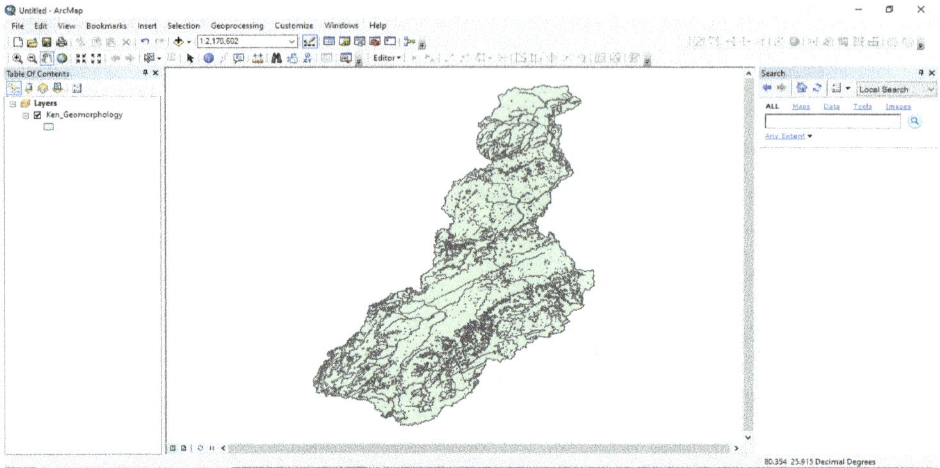

Right Click on the file name **Geomorphology_Ken** showing in Layers panel. Then, click on **Properties** option, and go to the **Symbology** option. After this, go to:

Categories > **Value Field** > **LEGEND_SHO** > **Add All Values** > **Apply** > **OK**

Different geomorphology units will be displayed in Layers panel.

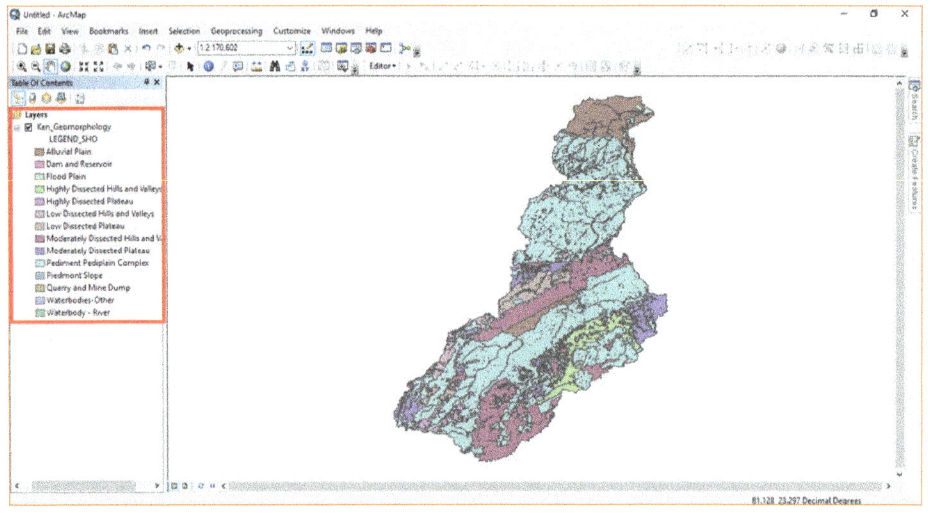

LINEAMENT DENSITY

Open the ArcMap application. Add the vector data of **Lineament Form** downloaded from the Bhukosh portal. Then, add the shape file of base map of Ken Basin.

Click on the **Search** option and type the **clip** and hit the **search** icon. Then, double click on the **Clip (Analysis)** tool in search result item.

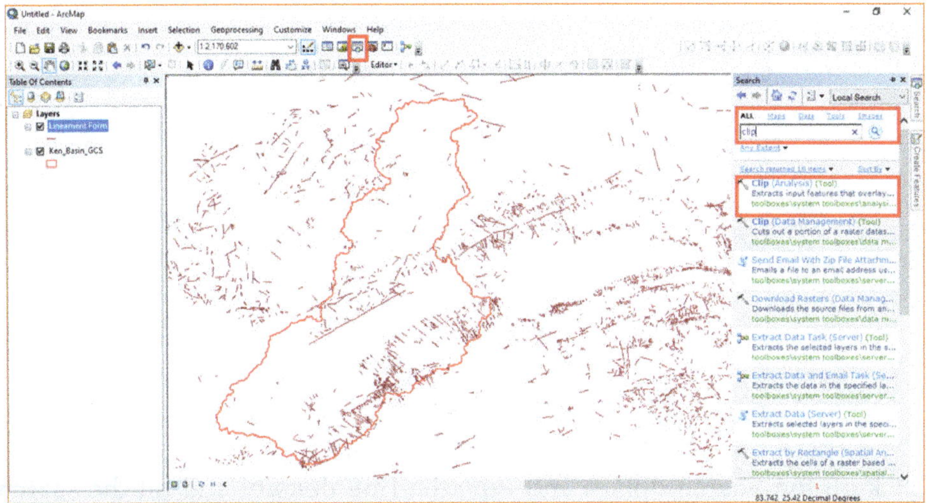

Add Lineament layer as an **Input Features**, select Boundary Shape file Ken_Basin_GCS in **Clip Features** option, and give the **Output Feature Class** as Lineament_Clip. Then, hit the **OK** Button.

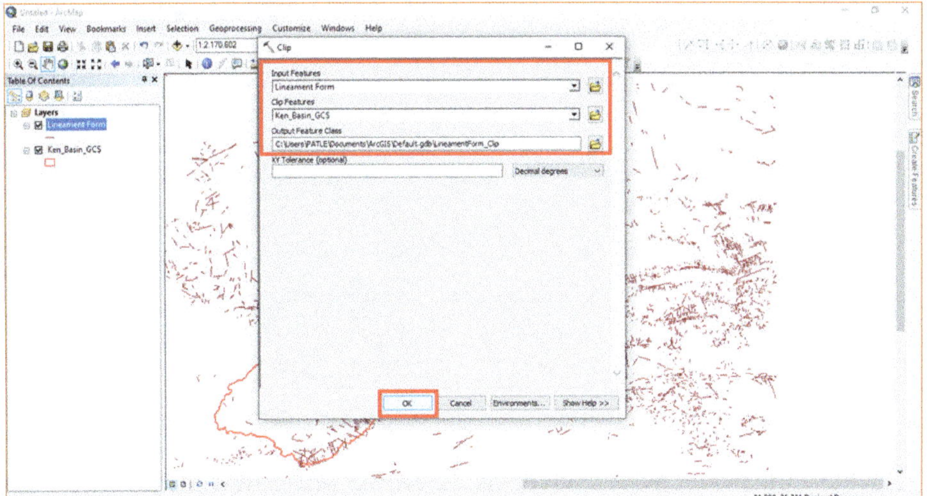

The Lineament layer has been clipped by Basin Shapefile.

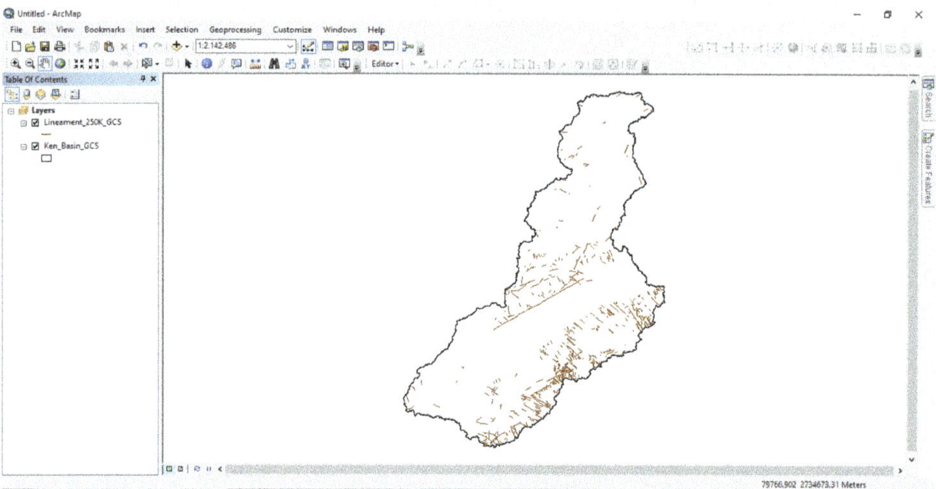

Click on the **Search** option and type the **line density** and hit the **search** icon. Then, double click on the **Line Density (Spatial Analyst)** tool in search result item.

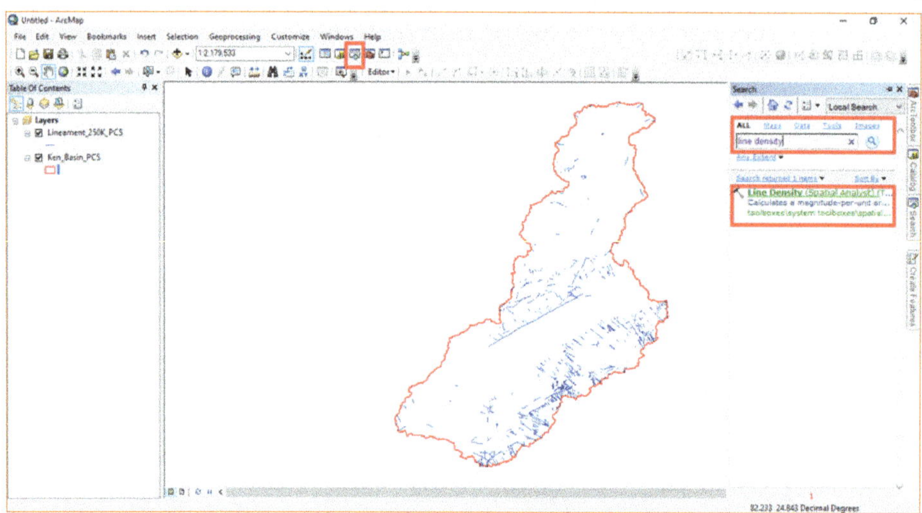

Add the Lineament vector layer as an **Input polyline features**, give the **output raster** as Lineament_Densitry.tif, Output Cell Size may be **10** or **30**, and select the **Area Unit** is SQUARE_KILOMETERS. Then, hit the **OK** Button.

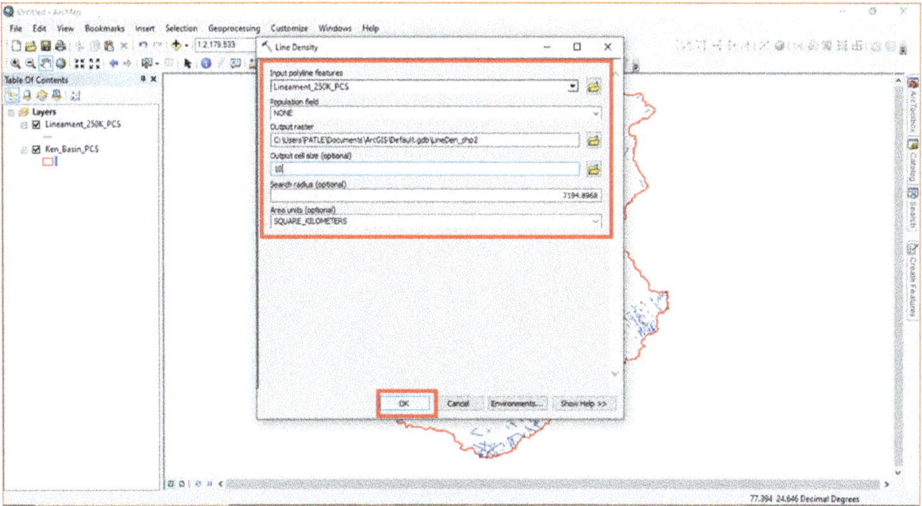

Lineament_Density.tif has been generated.

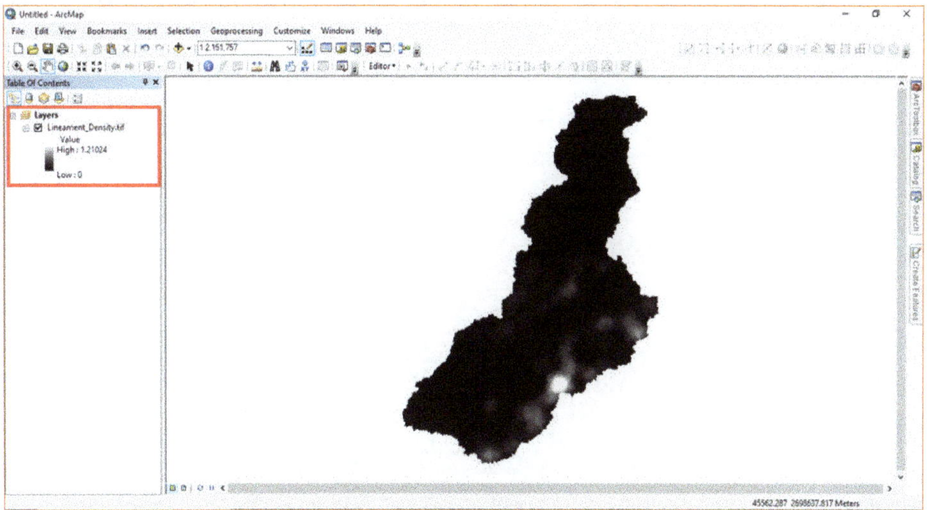

LAND USE/ LAND COVER

Open the ArcMap application. Add the raster data of **Land Use/ Land Cover (LULC)** downloaded from the Land Cover Explorer.

LULC consists of predefined values which represent different classes of land cover is given in link. (https://www.arcgis.com/home/item.html?id=cfcb7609de5f478eb7666240902d4d3d)

Then, add the shape file of base map of Ken Basin.

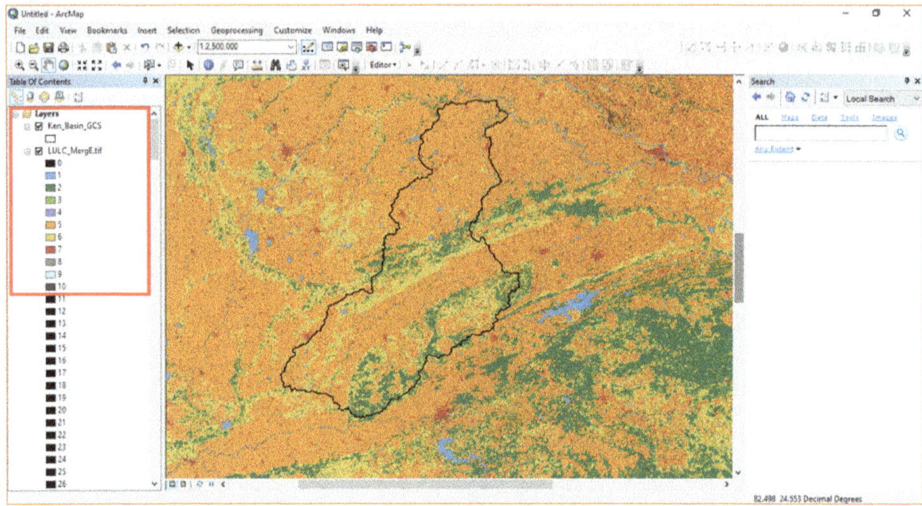

Click on the **Search** option and type the **raster clip** in search box and hit the **search** icon. Then, double click on the **Clip (Data Management)** tool in search result item.

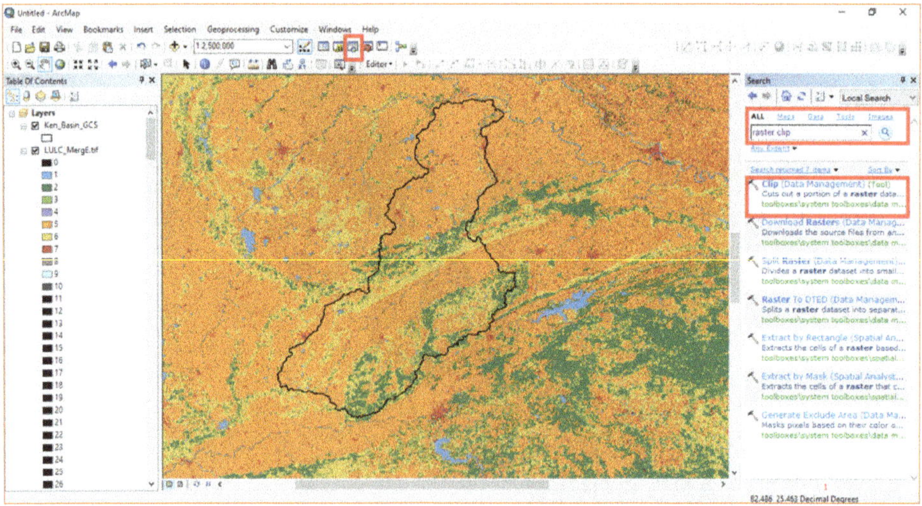

Add LULC_Merge.tif in **Input Raster** option, select Ken_Basin_GCS.shp vector file in **Output Extent option**, Tick mark the Check Box (**Use In put Feature for Clipping Geometry**), and give the output name in the **Output Raster Dataset** option. **NoData Value** should be 0. Click on the **OK** button.

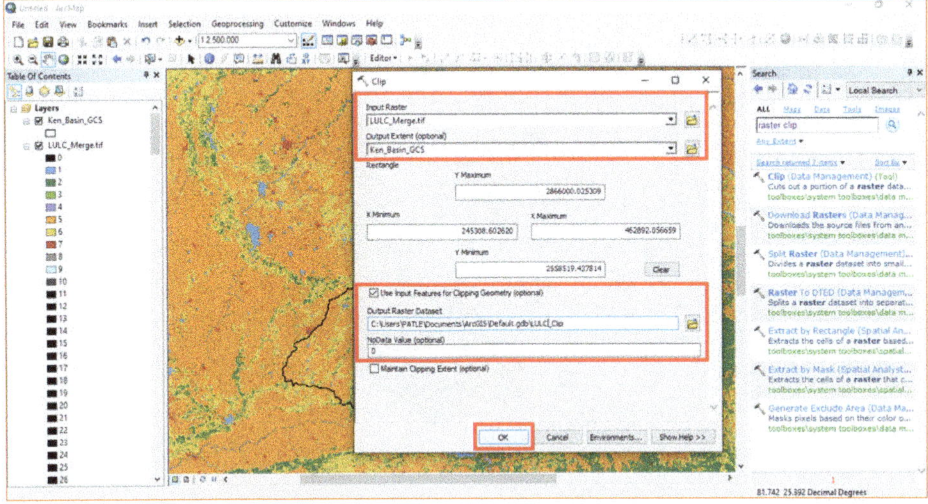

Land use/ land cover (LULC) of Ken River Basin has been clipped. Classes of LULC has been renamed according to Land Cover Portal.

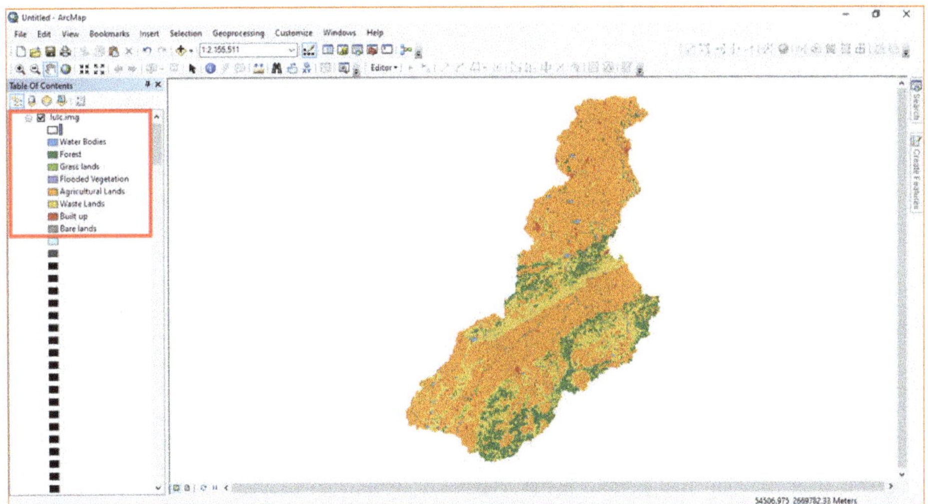

SOIL

In this study, the of Soil map of the Ken Map has been developed using the digitization of soil map acquired from the literature. This map also been developed using the soil map sheets collected from the NBSS&LUP, Nagpur. Firstly, georeferenced this map using the vector file of Ken Basin and then digitized it.

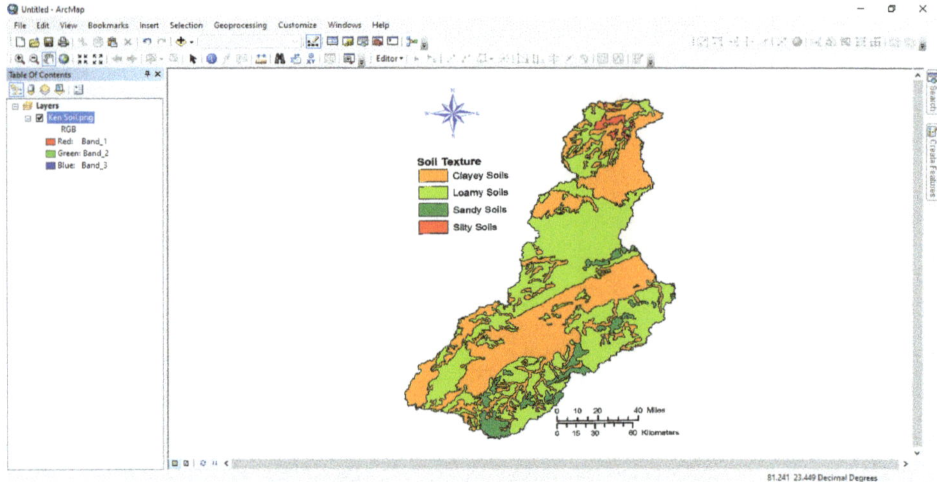

Open the ArcMap application. Add the vector data of **Ken_Soil.shp**.

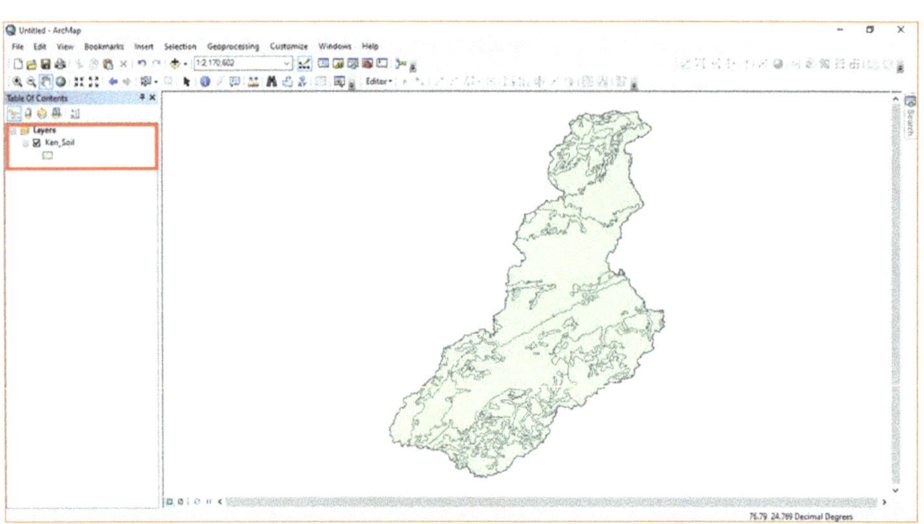

Right Click on the file name **Ken_Soil** showing in Layers panel. Then, click on **Properties** option, and go to the **Symbology** option. After this, go to:

Categories > **Value Field** > **Texture** > **Add All Values** > **Apply** > **OK**

Different soil texture will be displayed in Layers panel.

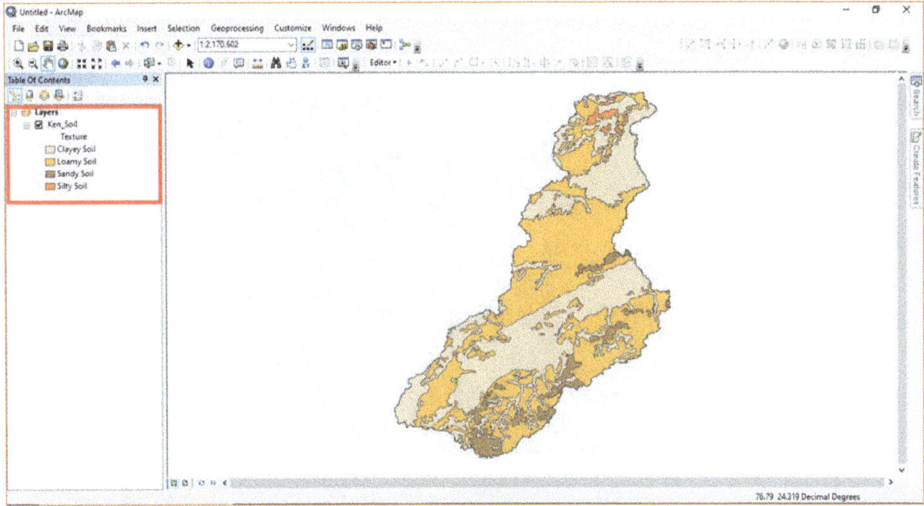

SLOPE

Open the ArcMap application. Add the raster data of **Digital Elevation Model (DEM)**. Then, add the shape file of base map of Ken Basin.

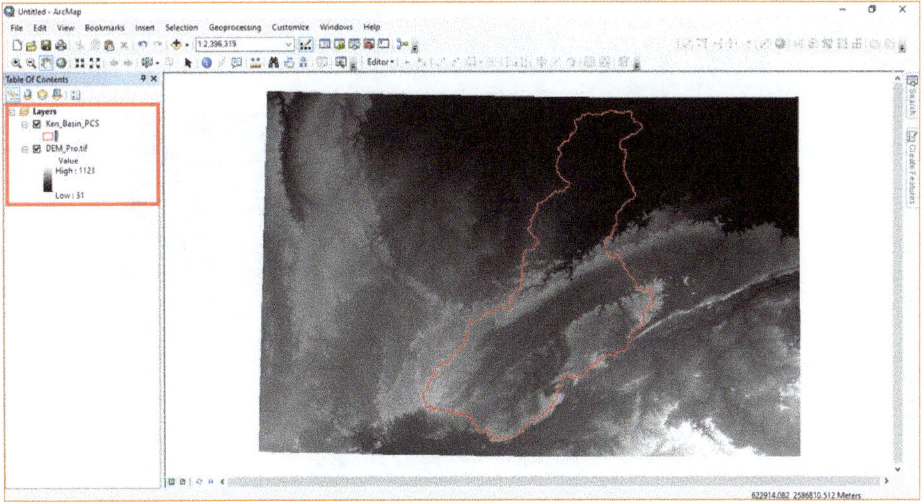

Click on the **Search** option and type the **raster clip** and hit the **search** icon. Then, double click on the **Clip (Data Management)** tool in search result item.

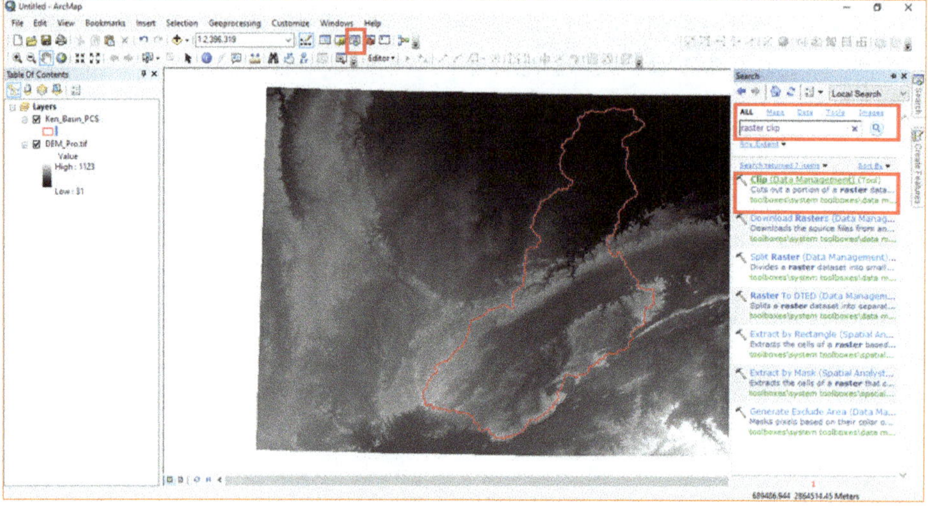

Add DEM_Pro.tif in **Input Raster** option, select Ken_Basin_PCS.shp vector file in **Output Extent** option, Tick mark the Check Box (**Use In put Feature for Clipping Geometry**), and give the output name in the **Output Raster Dataset** option. **No Data Value** should be 0. Click on the OK button.

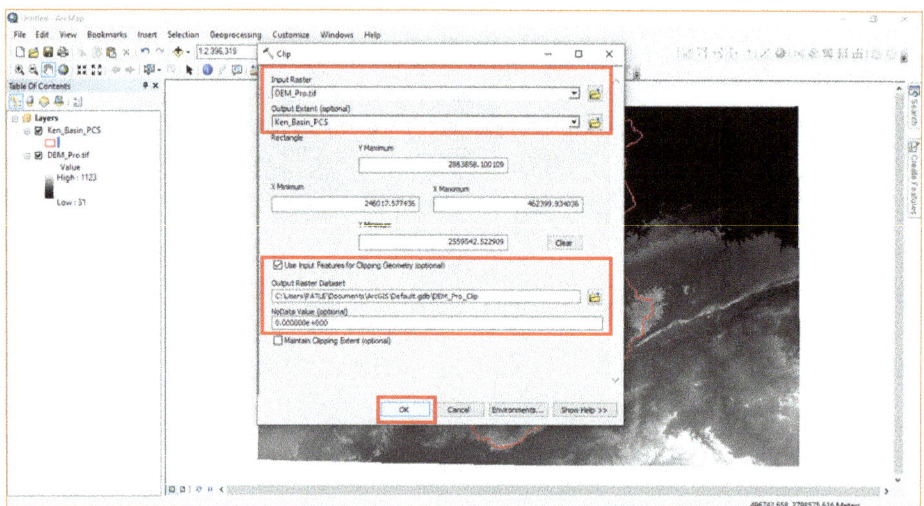

Digital Elevation Model (DEM) of Ken Basin has been clipped.

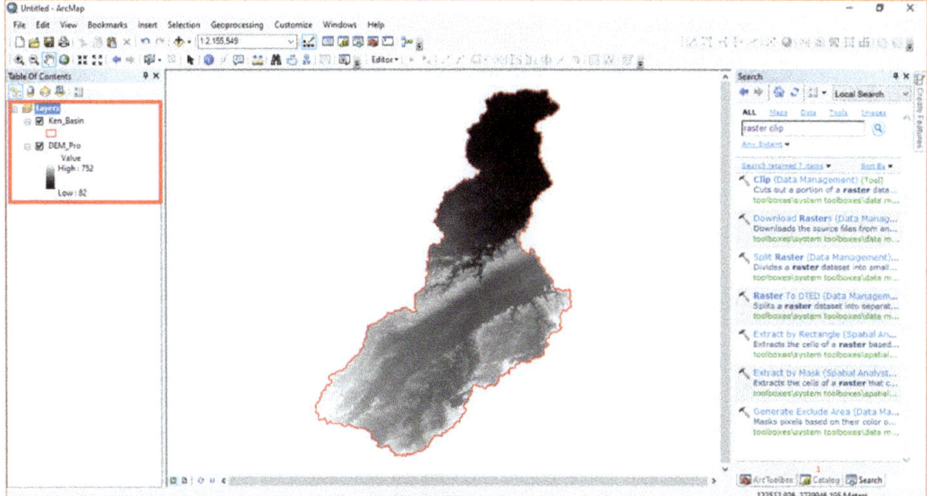

Click on the **Search** option and type the **Fill** and hit the **search** icon. Then, double click on the **Fill (Spatial Analyst)** tool in search result item.

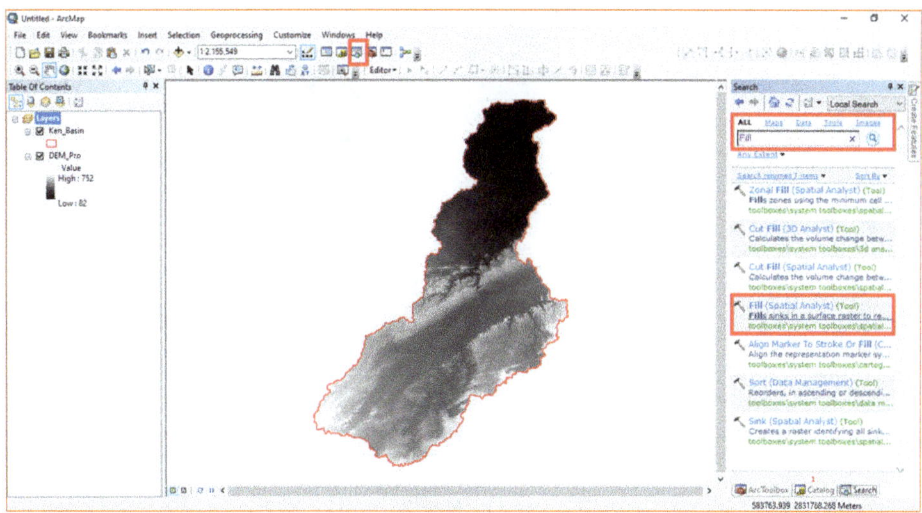

Select DEM_Pro in **Input surface raster**, give the output name Fill_Ken in **Output surface raster** option, and hit the **OK** button.

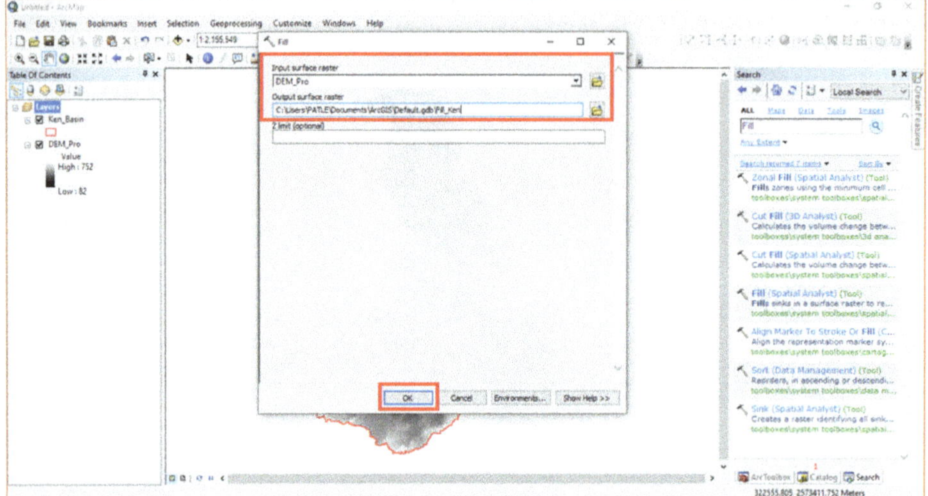

Fill Ken raster data has been generated.

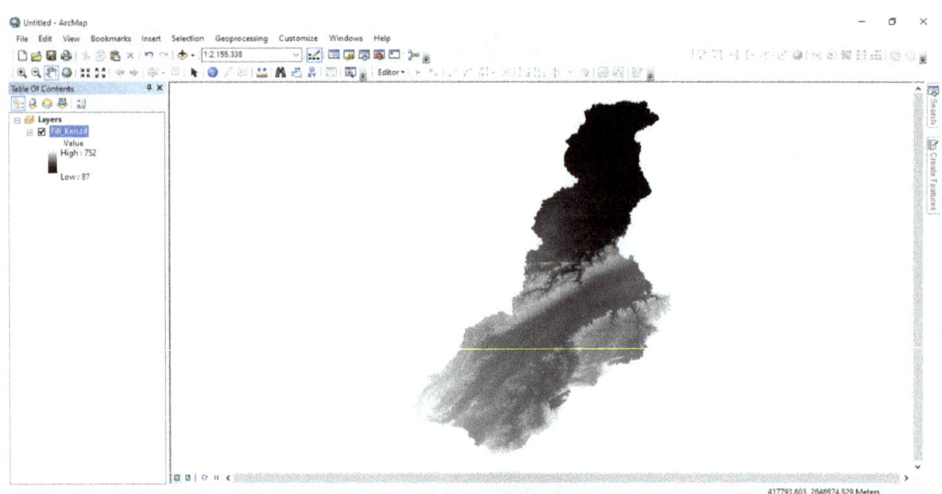

Click on the **Search** option and type the **Slope** and hit the **search** icon. Then, double click on the **Slope (Spatial Analyst)** tool in search result item.

Select Fill_Ken in **Input raster**, give the output name Slope_Percent.tif in **Output raster** option, select the PERCENT_RISE in **Output measurement option**, and hit the **OK** button.

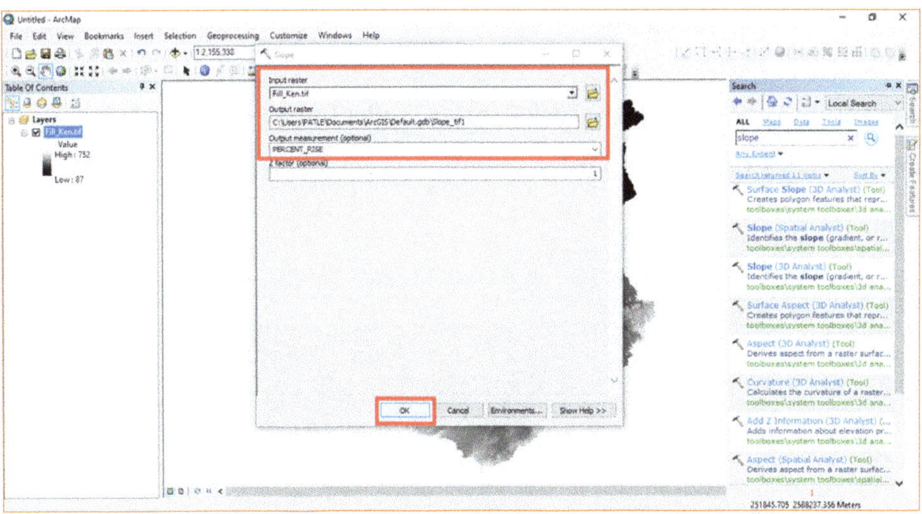

Slope map of Ken Basin in Percent has been generated.

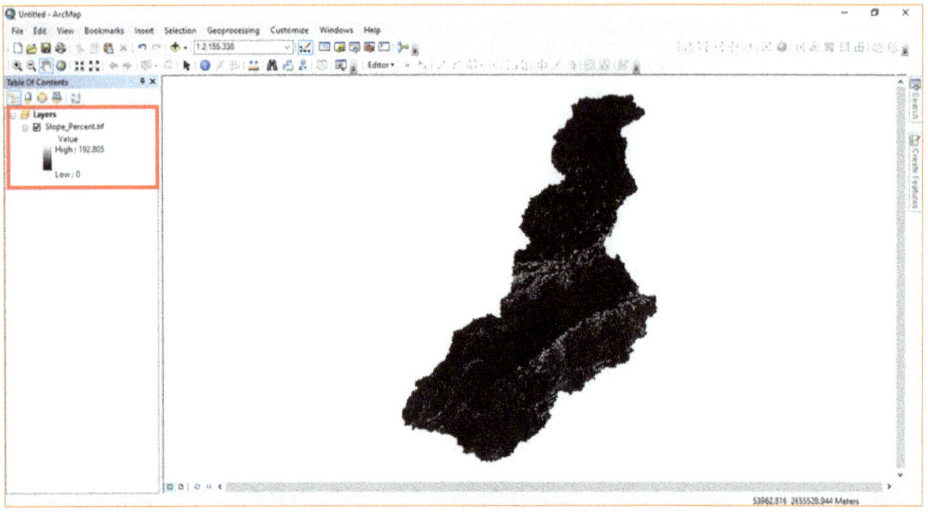

DRAINAGE DENSITY

In order to develop the drainage density, it is need to have '**flow direction**' and '**condition data**'. These data were developed during the base map preparation.

Go to:

ArcTool Box > Spatial Analyst Tools > Hydrology > Stream Order

Double Click on **Stream Order** option.

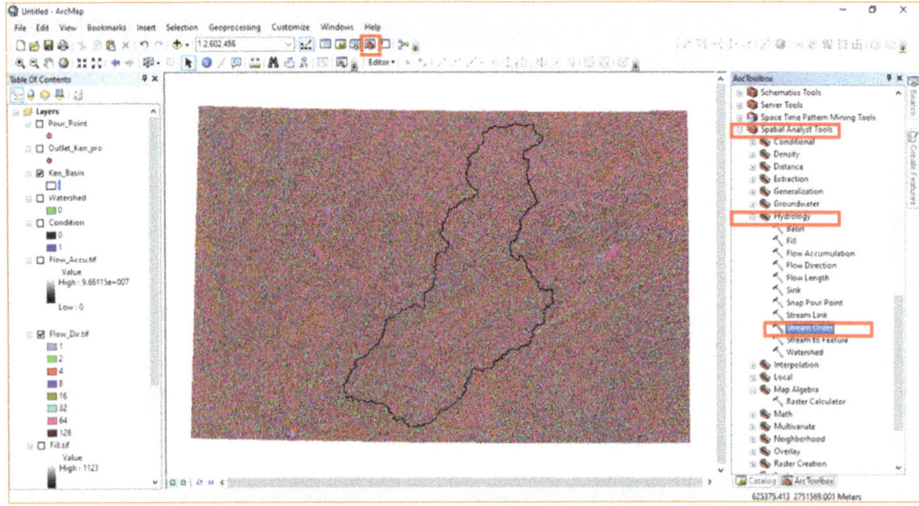

Condition.tif raster select in **Input stream raster** option, select Flow_dir.tif file in **Input flow direction** raster option, and give the output name as stream_odr.tif in **output raster** option. Select STRAHLER option in **Method of stream ordering** and click on **OK** Button.

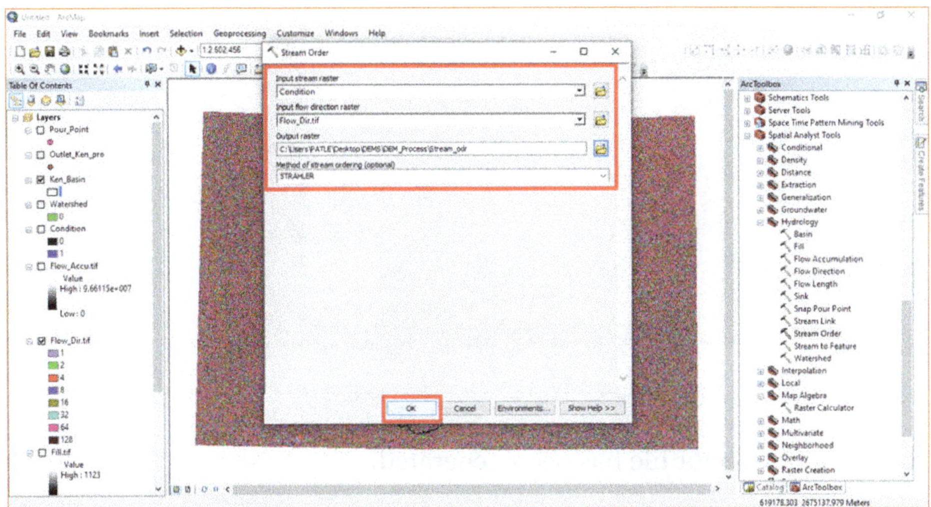

The Raster file of stream order has been generated. Go to:

ArcTool Box > Spatial Analyst Tools > Hydrology > Stream to Feature. Double-Click on **Stream to Feature** option.

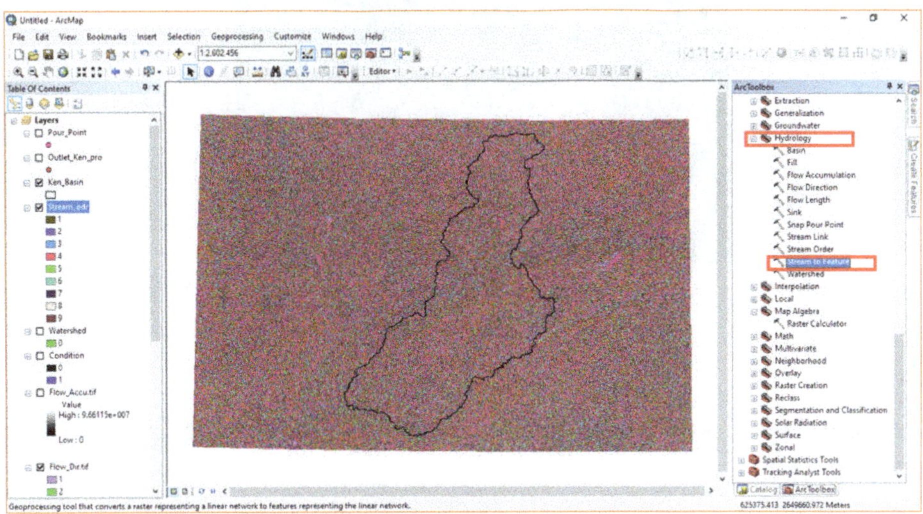

Select Sream_odr.tif file as **Input stream raster**, choose Flow_Dir.tif file as **Input flow direction raster**, give the name of **Output polyline feature** as Stream_Order.shp, and hit the **OK** button.

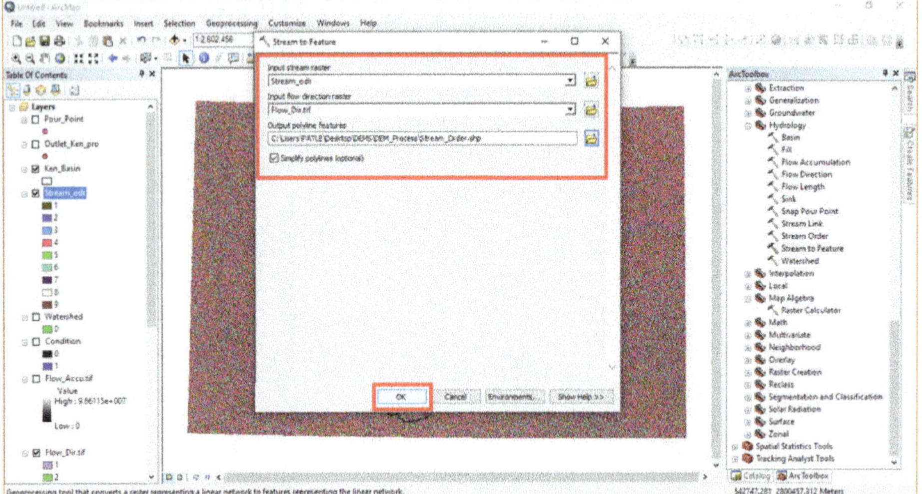

Stream Order vector file has been generated.

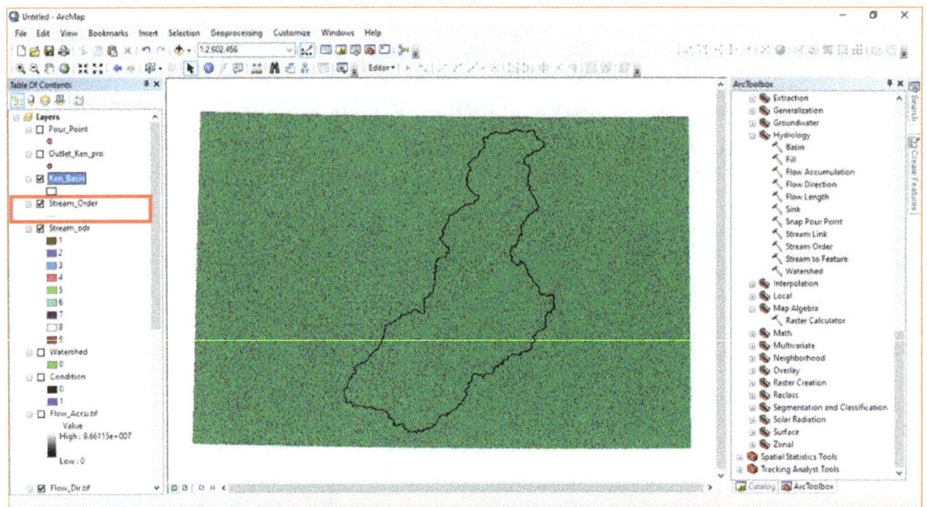

Go to:

ArcTool Box > **Analysis Tools** > **Extract** > **Clip.** Double-Click on **Clip** option.

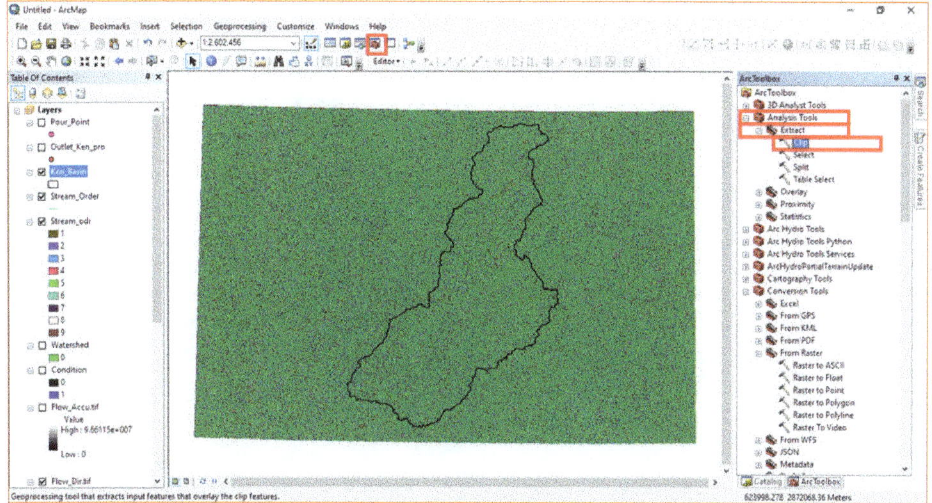

Add Stream_Order as an **Input Features**, select Boundary Shape file Ken_Basin in **Clip Features** option, and give the **Output Feature Class** as Stream_Order_Clip. Then, hit the **OK** Button.

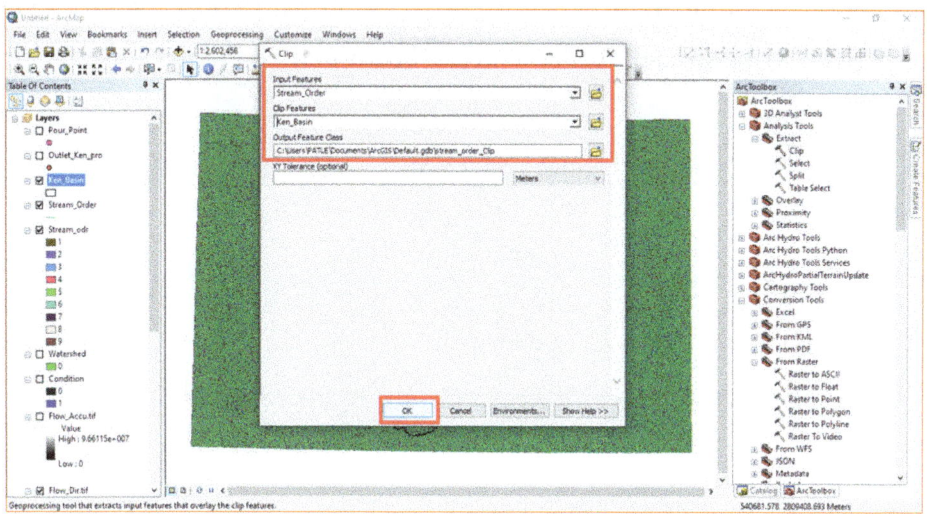

Stream Order of Ken Basin has been clipped.

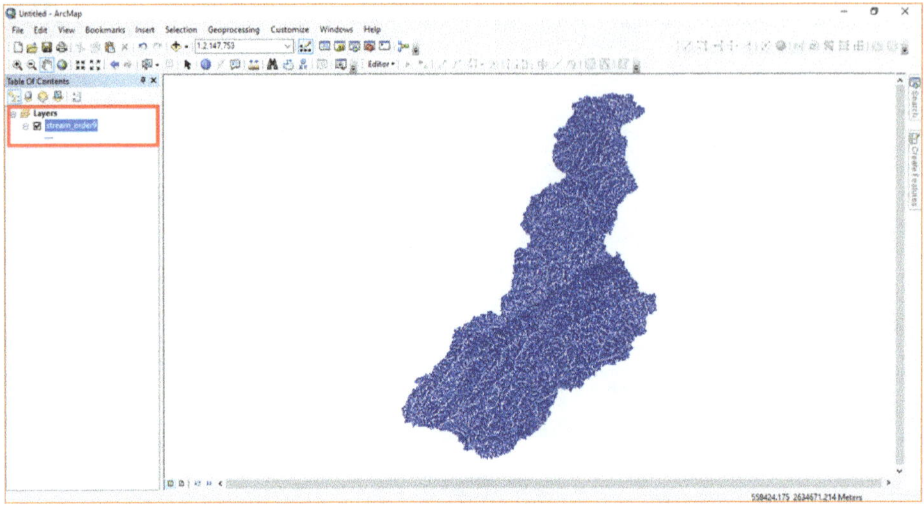

Right Click on the file name **Stream_Order9** showing in Layers panel. Then, click on **Properties** option, and go to the **Symbology** option. After this, go to:

Categories > **Value Field** > **GRID_CODE** > **Add All Values** > **Apply** > **OK**

All the order of stream available in the basin will be displayed in Layers panel.

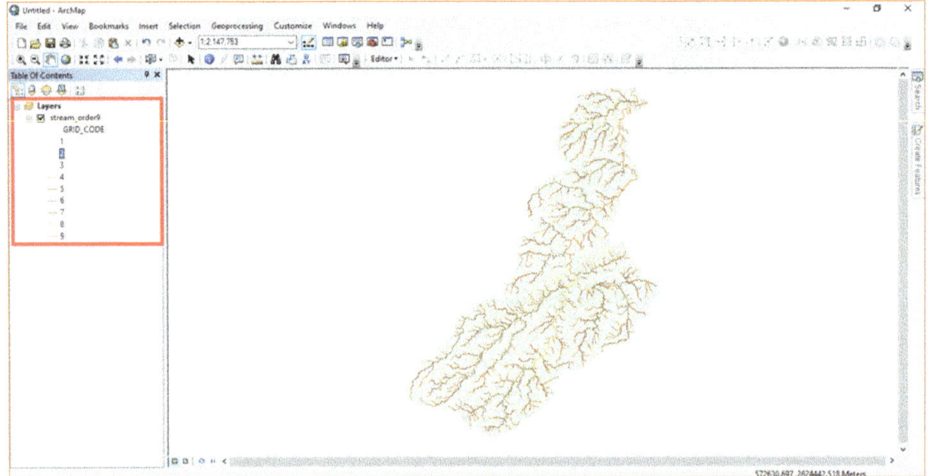

Click on the **Search** option and type the **line density** and hit the **search** icon. Then, double click on the **Line Density (Spatial Analyst)** tool in search result item.

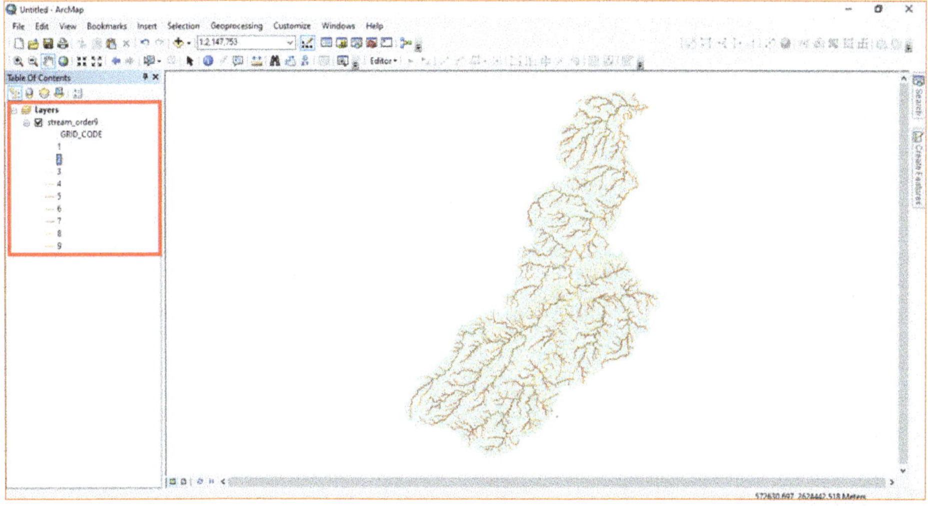

Add the stream order vector layer as an **Input polyline features**, give the **output raster** as Drainage_Densitry.tif, Output Cell Size may be **10** or **30**, and select SQUARE_KILOMETERS in **Area Unit** option. Then, hit the **OK** Button.

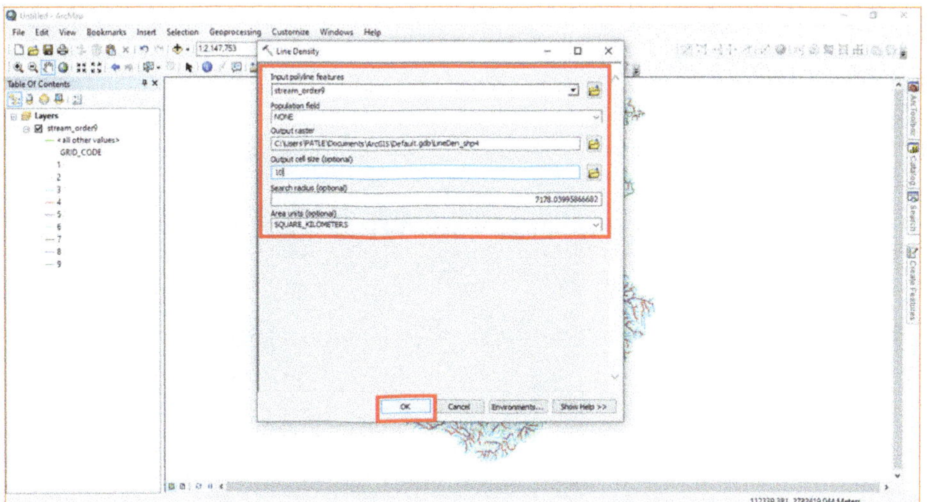

Drainage Density of the Ken Basin has been generated.

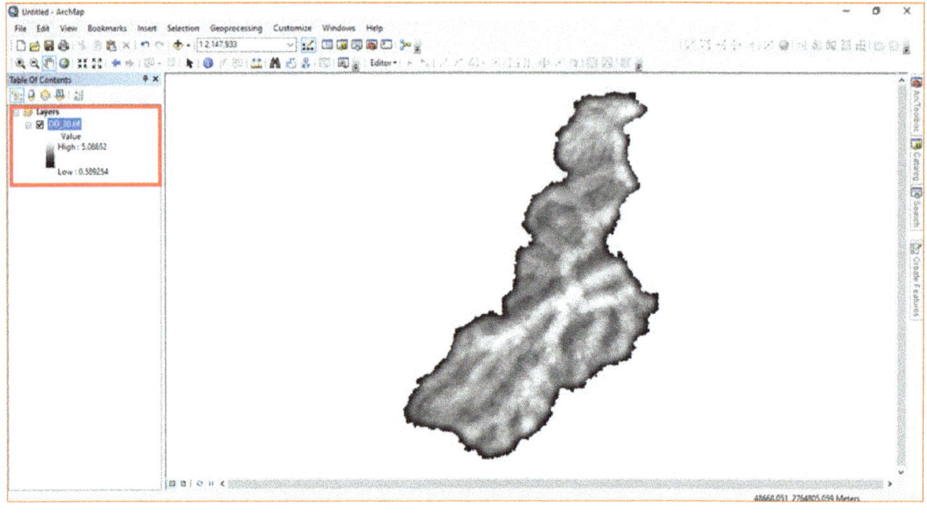

RAINFALL

Firstly, downloaded rainfall data is converted from grid file to vector file format like .shp using IMD Data Converter. After the conversion of data, the point shape file of rainfall is added in ArcMap application of ArcGIS software.

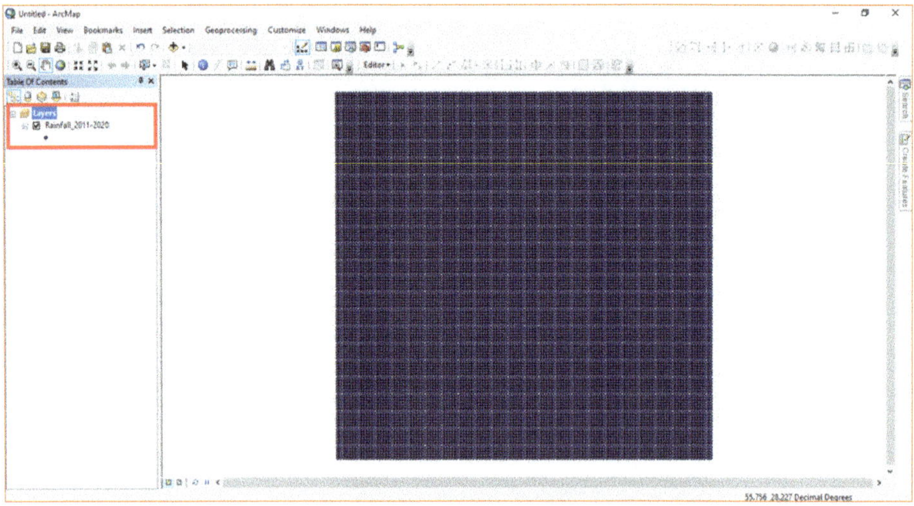

Add the Base map of Ken River Basin over the Point data.

Click on the **Search** option and type the **clip** and hit the **search** icon. Then, double click on the **Clip (Analysis)** tool in search result item.

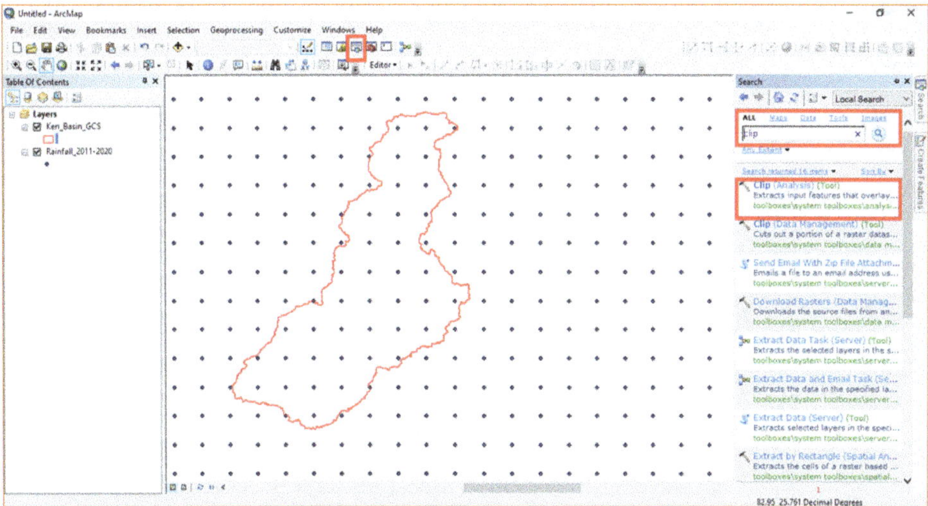

Add 'Rainfall_2011-2020' as an **Input Features**, select Boundary Shape file 'Ken_Basin' as a **Clip Features**, and give the **Output Feature Class** as Rainfall_Clip. Then, hit the **OK** Button.

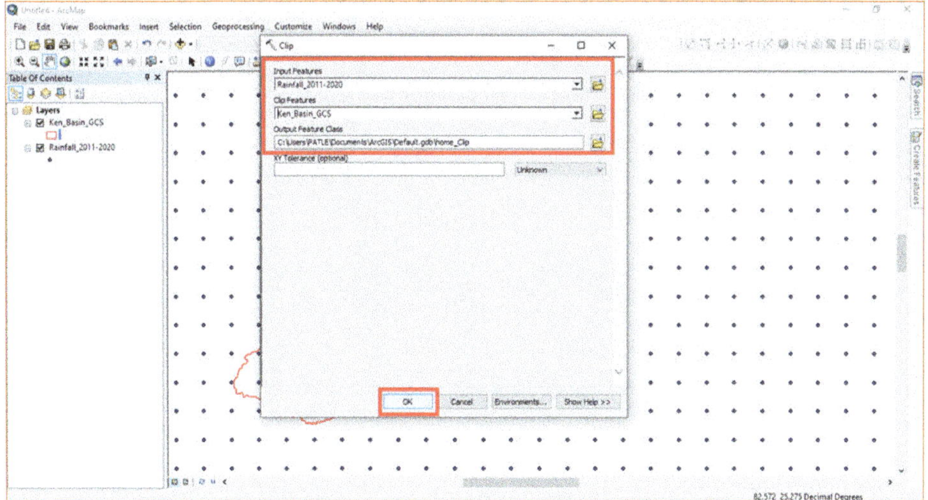

Rainfall Point data of Ken River Basin has been clipped.

Click on the **Search** option and type the **idw** and hit the **search** icon. Then, double click on the **IDW (Spatial Analyst)** tool in search result item.

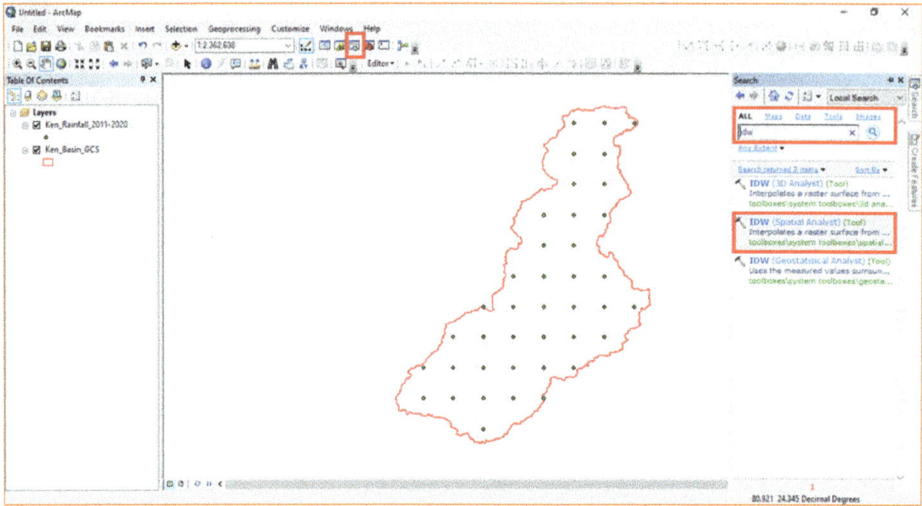

Select the Rainfall of Ken Basin as an **Input point features**, give the **Z value field** such as Rainfall Value (in mm) option, provide the output name Rainfall_Ken.tif in **Output raster** option, assign the cell size 10 or 30 in **output cell size** option, and click on **OK** option.

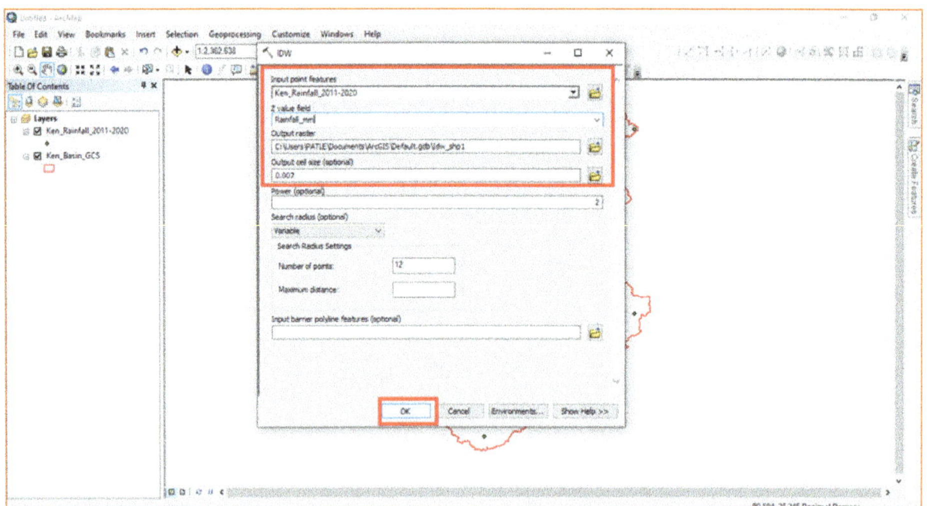

Raster map of Rainfall (2011-2020) of Ken Basin has been developed.

It is the Monsoon Rainfall, which ranges from 595 mm to 1226 mm over the Ken River Basin.

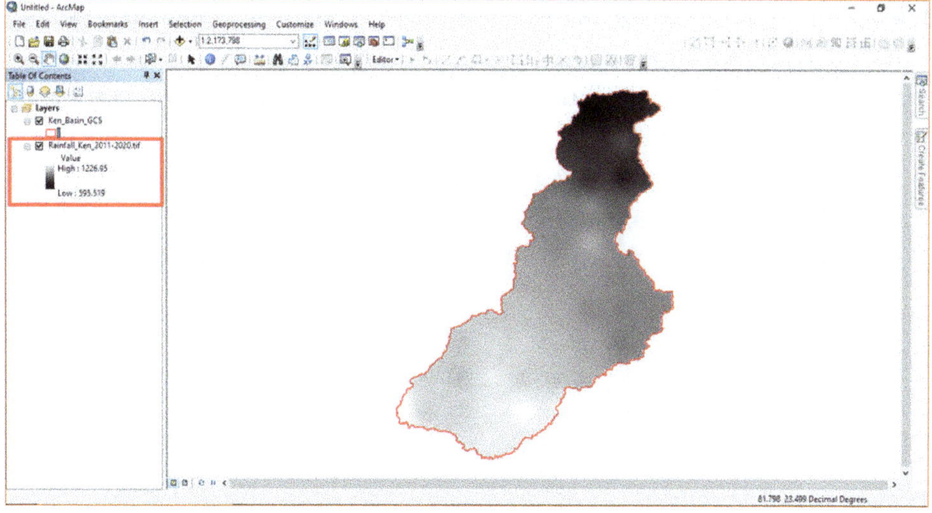

RASTER CONVERSION OF THEMATIC MAPS

Out of eight themes, the three themes namely geology, geomorphology and Soil are in the vector format. So, all these themes need to convert into raster format.

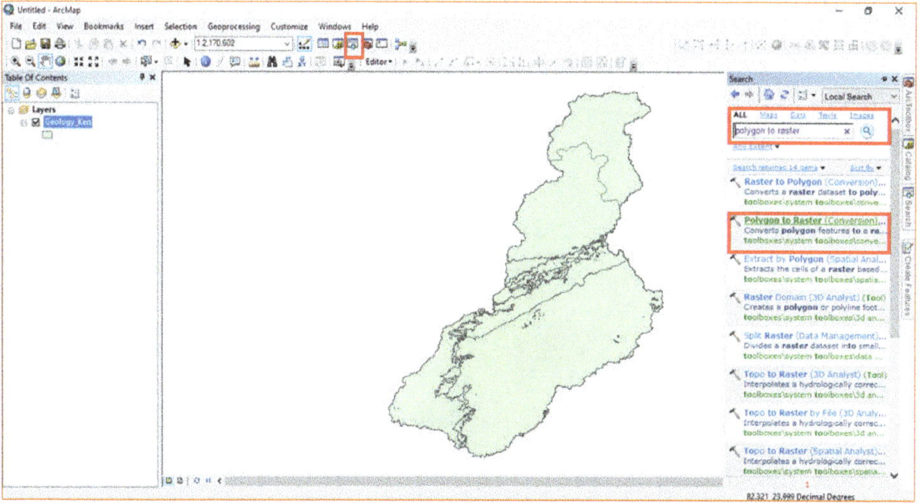

Firstly, Open the ArcMap application and add the vector data of Geology.

Click on the **Search** option and type the **polygon to raster** in search box and hit the **search** icon. Then, double click on the **Polygon to Raster** conversion tool in search result item.

Select Geology_Ken in **Input Features** option, choose GROUP_Name in **Value Field** option, give the output name Geology.tif in **Ouput Raster Dataset** option, assign cell size 30, and click on the **OK** Button.

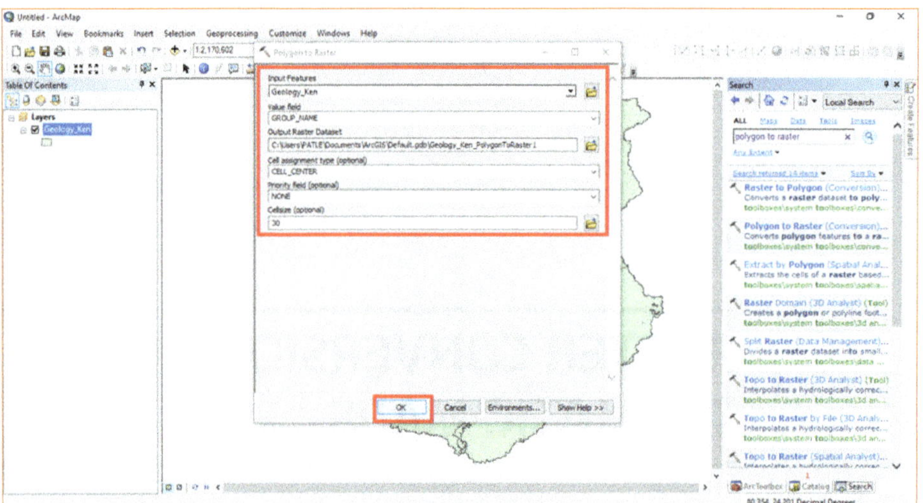

Raster of Geology of Ken Basin has been developed.

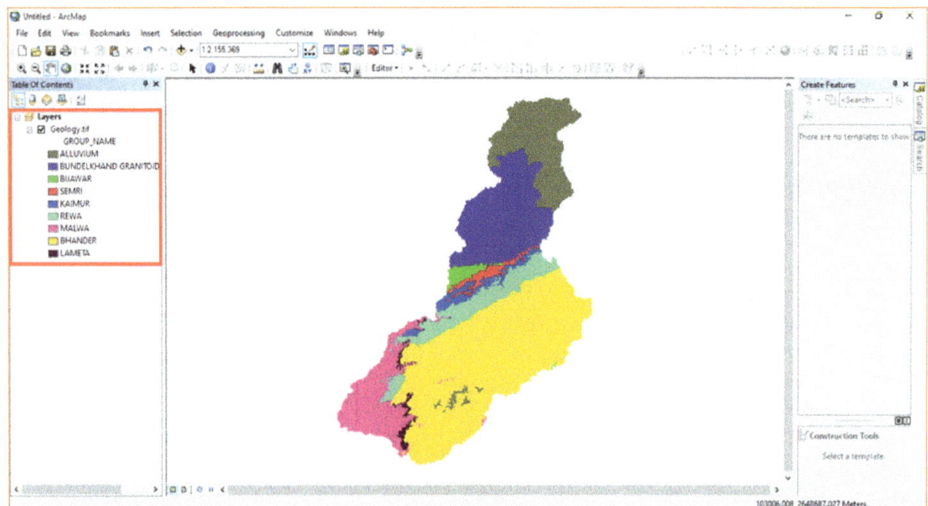

Open the ArcMap application and add the vector data of Geomorphology.

Click on the **Search** option and type the **polygon to raster** in search box and hit the **search** icon. Then, double click on the **Polygon to Raster** conversion tool in search result item.

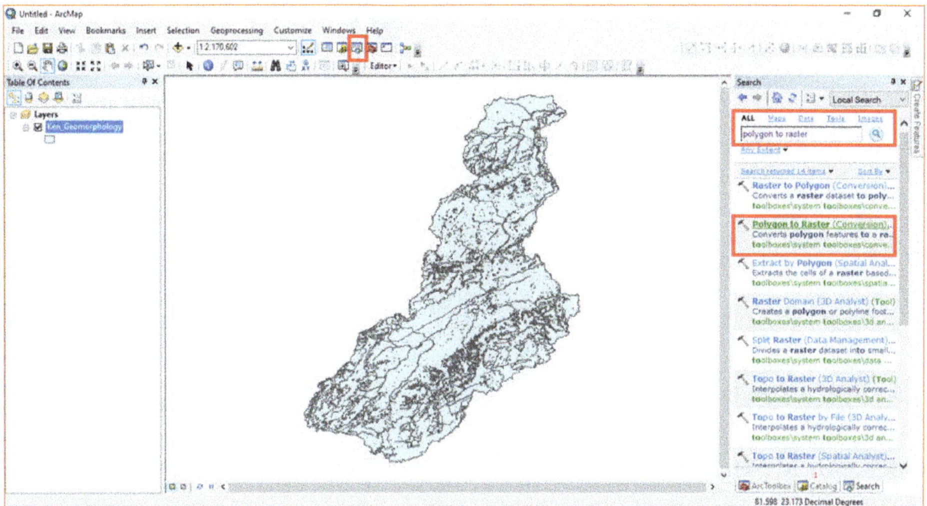

Select Geomorphology_Ken in **Input Features** option, choose LEGEND_SHO in **Value Field** option, give the output name Geomorphology.tif in **Output Raster Dataset** option, assign cell size 30, and click on the **OK** Button.

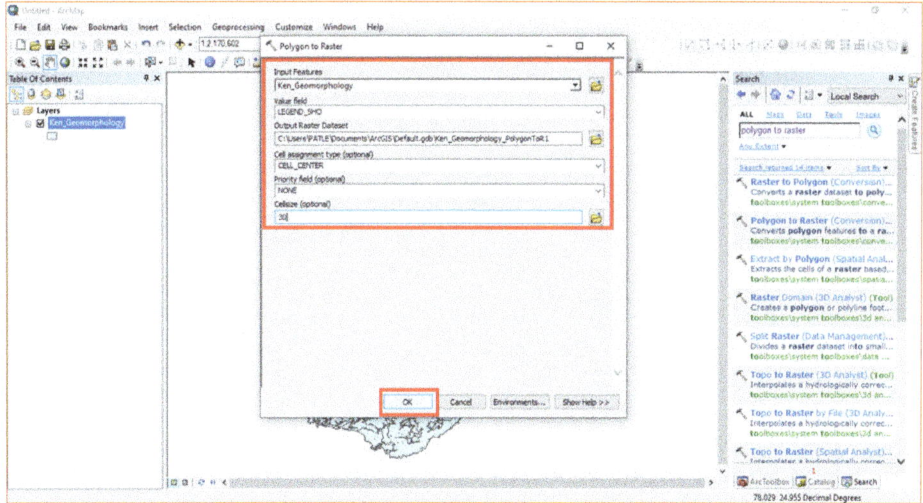

Raster data of Geomorphology of Ken Basin has been developed and showing different geomorphological units in **Layers** panel.

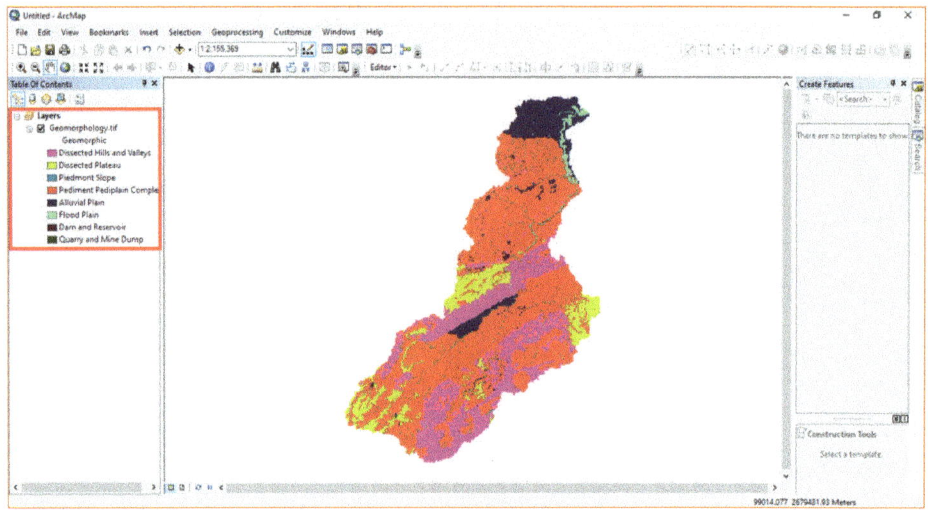

Open the ArcMap application and add the vector data of soil.

Click on the **Search** option and type the **polygon to raster** in search box and hit the **search** icon. Then, double click on the **Polygon to Raster** conversion tool in search result item.

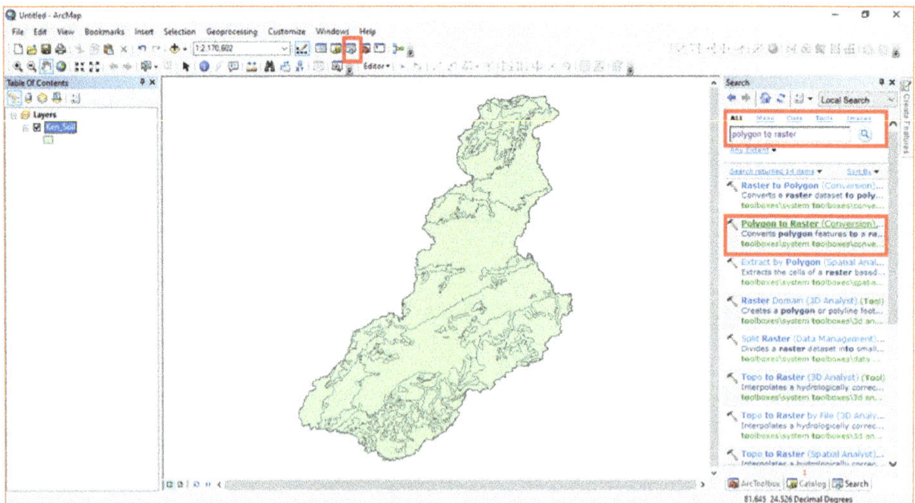

Select Ken_Soil in **Input Features** option, choose Texture in **Value Field** option, give the output name Soil_Texture.tif in **Output Raster Dataset** option, assign cell size 30, and click on the **OK** Button.

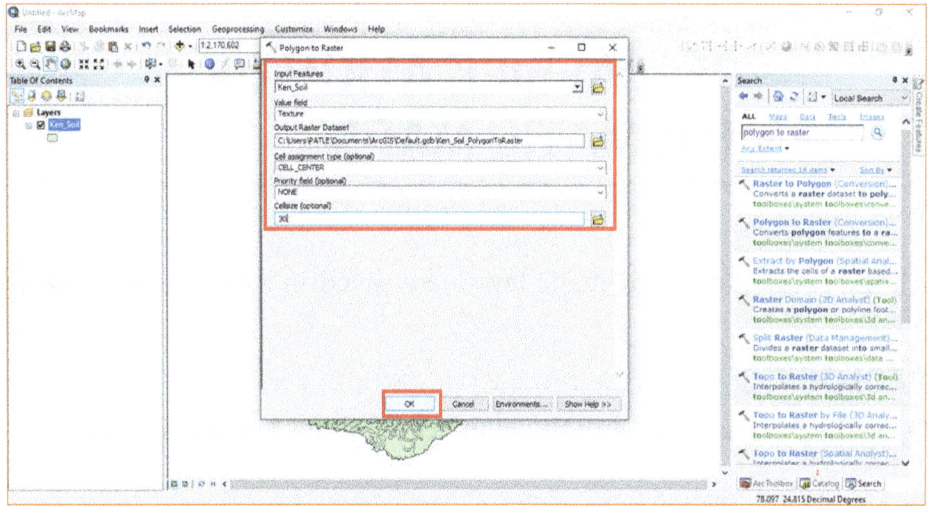

Raster data of Soil of Ken Basin has been developed and showing different soil texture in **Layers** panel.

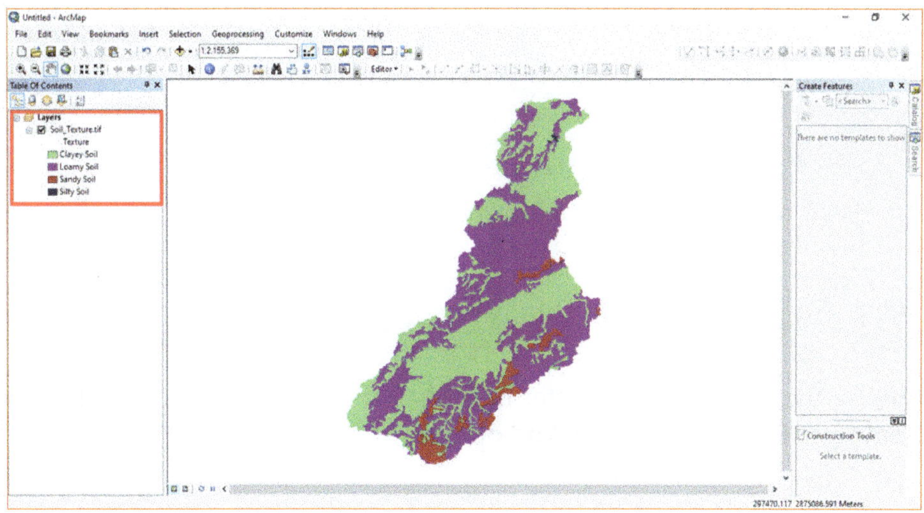

RECLASSIFICATION OF THEMATIC MAPS

All the raster data of thematic layers are saved in a folder name 'Raster Maps'.

GEOLOGY

Open the **ArcMap.exe** application and click on the **Add Data** option.

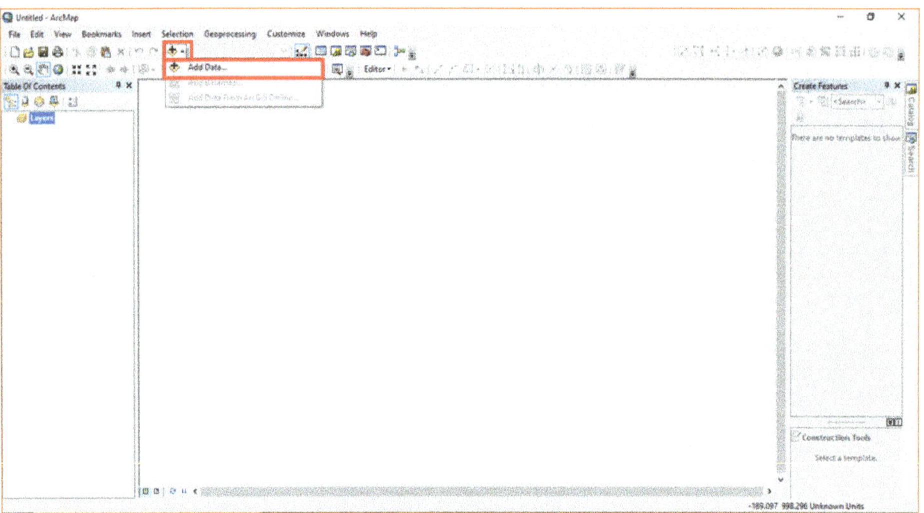

Browse the folder having raster files of all the thematic factors. Select the **Geology.tif** file and click on **Add** option.

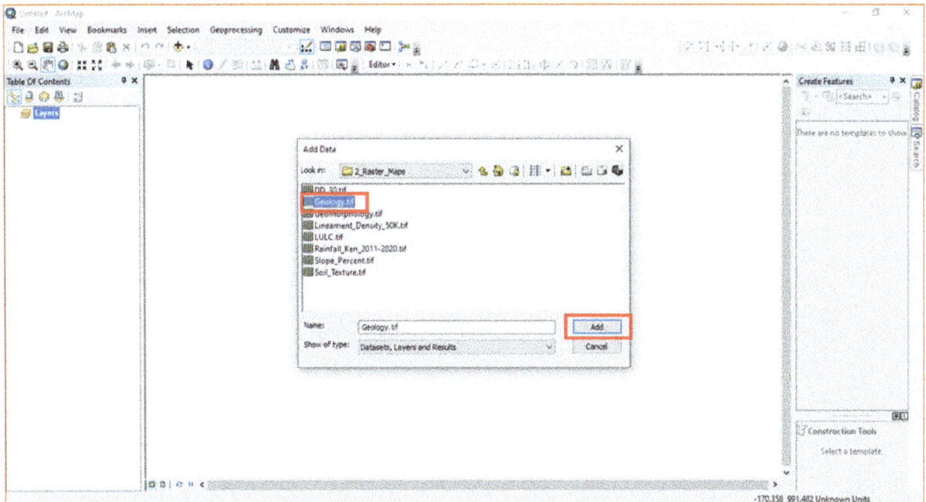

Geology map of Ken Basin is classified into different geological major groups.

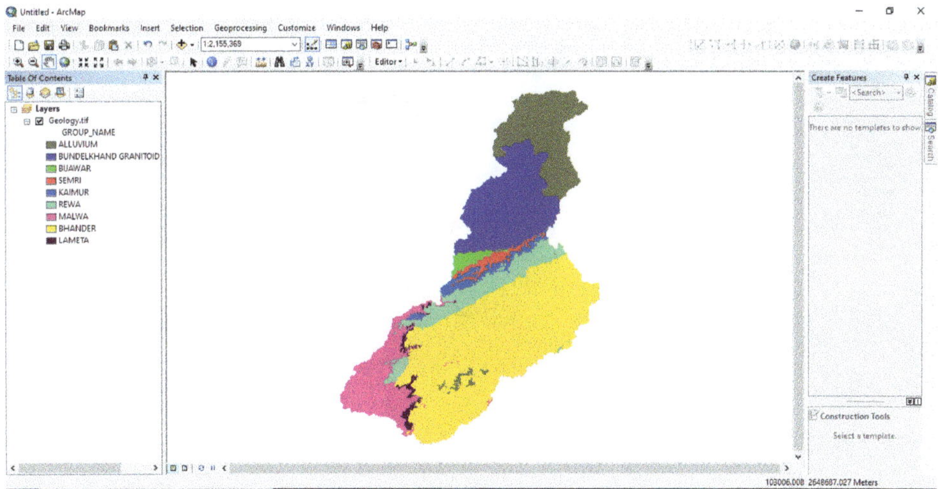

GEOMORPHOLOGY

Browse the 'Raster_Maps' folder, select on the **Geomorphology.tif** file and click on **Add** option.

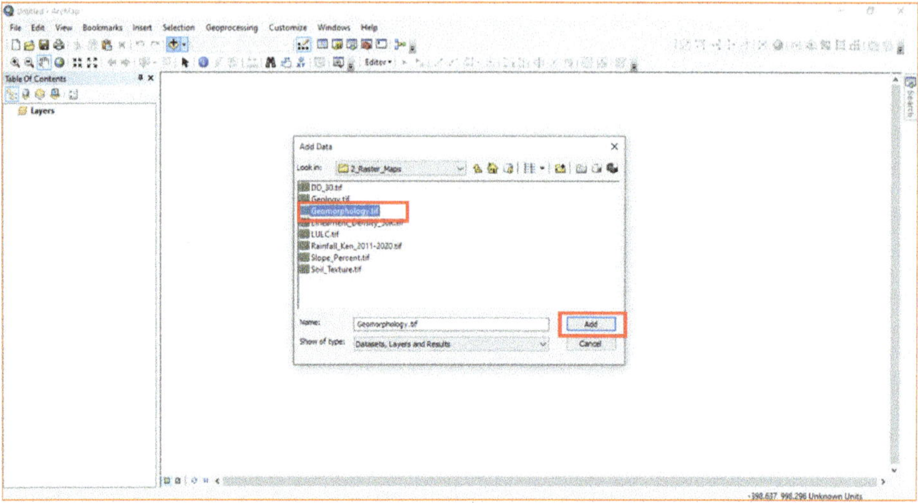

Geomorphology map of Ken Basin is classified into different geomorphological units.

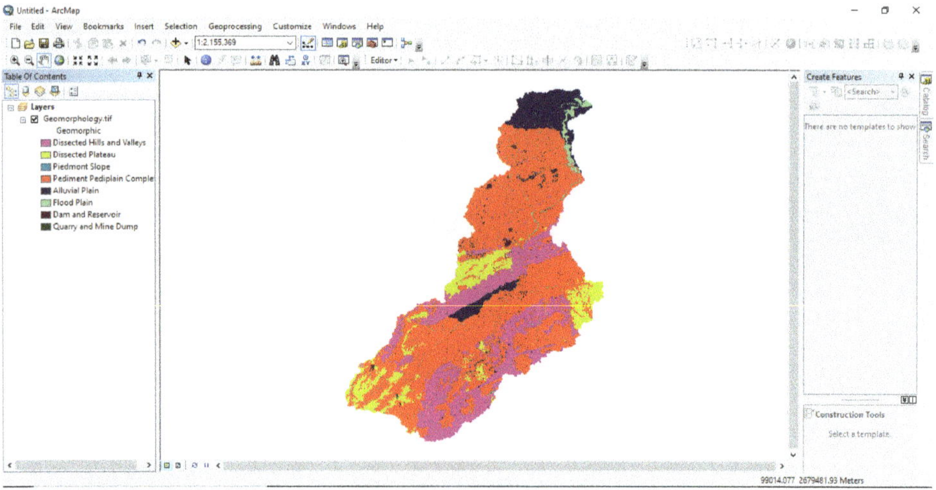

LINEAMENT DENSITY

Browse the 'Raster_Maps' folder, select on the **Lineament_Density.tif** file and click on **Add** option.

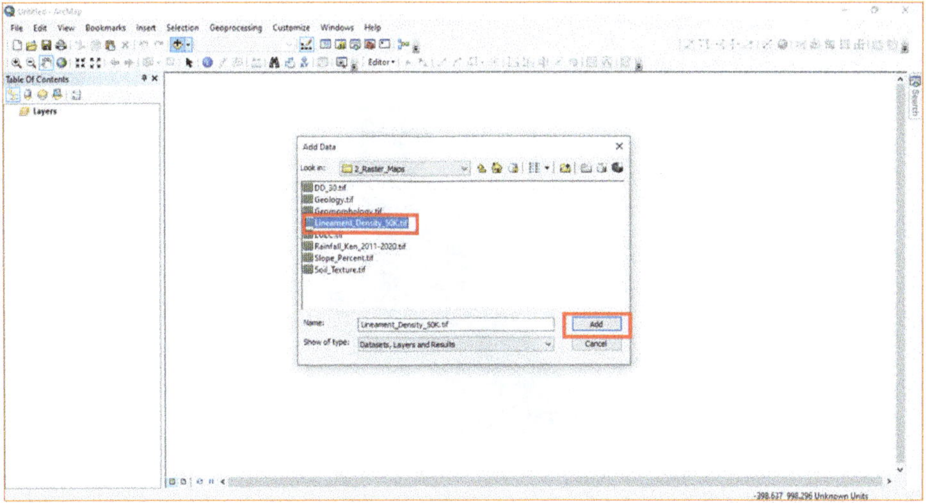

Lineament Density map is available in dynamic composite map form. It needs to be reclassified.

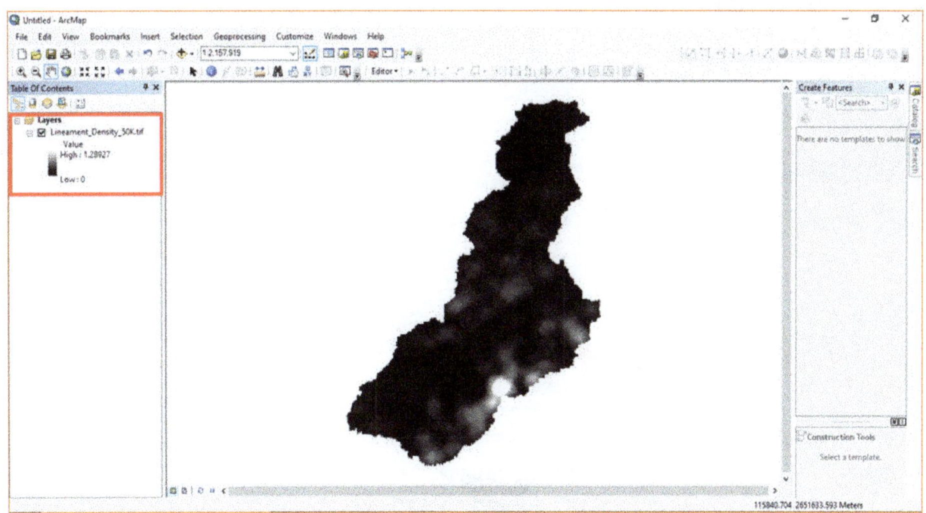

Right click on **Lineament_Density.tif** file showing in Layer panel. Click on **Properties** option.

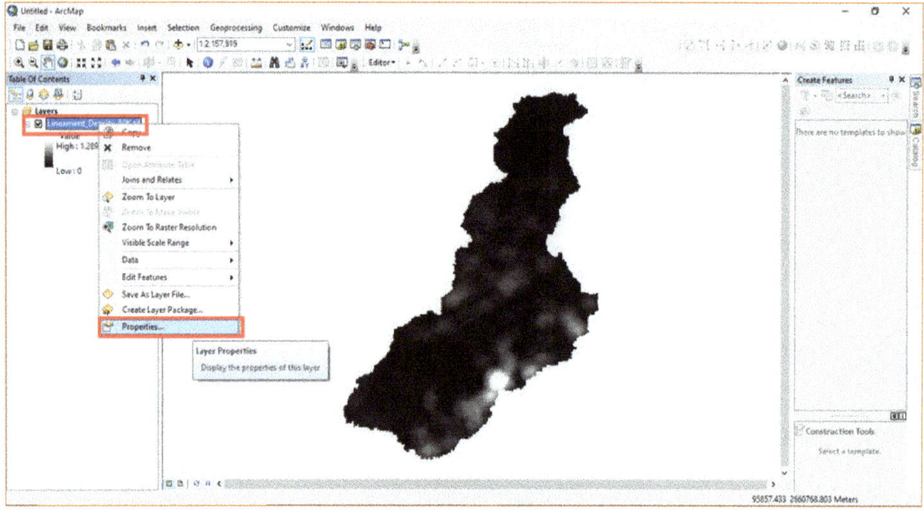

Click on **Symbology** option, select the **Classified** option, give the number of **Classes**, change the color from **Color Ramp**, click on **Apply** option, and hit the **OK** option.

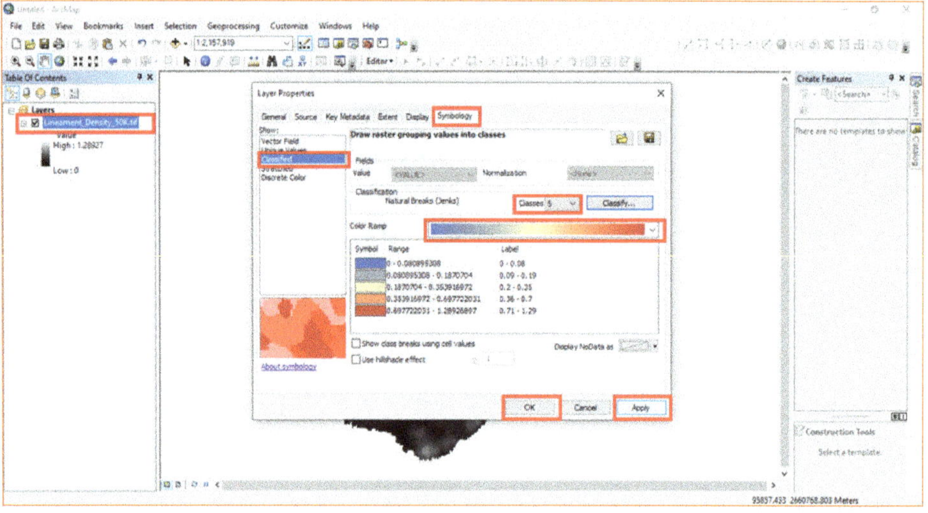

The lineament density of Ken basin has been classified into 5 classes using the **Natural Break (Jenks)** classification method. It needs to be **reclassified.**

Classification method may be changed according to the purpose of the study.

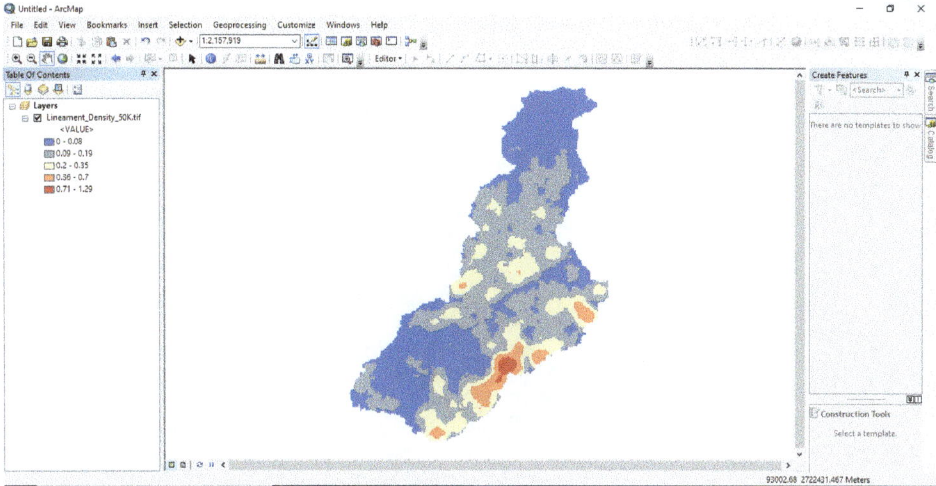

Click on **ArcTool Box** icon.

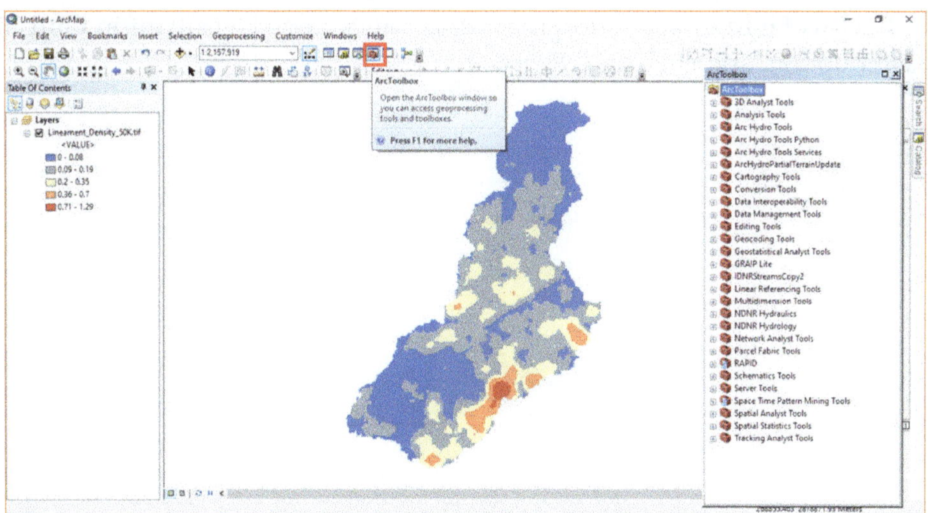

Go to:

ArcTool Box > Spatial Analyst Tools > Reclass > Reclassify. Double-Click on Reclassify tool.

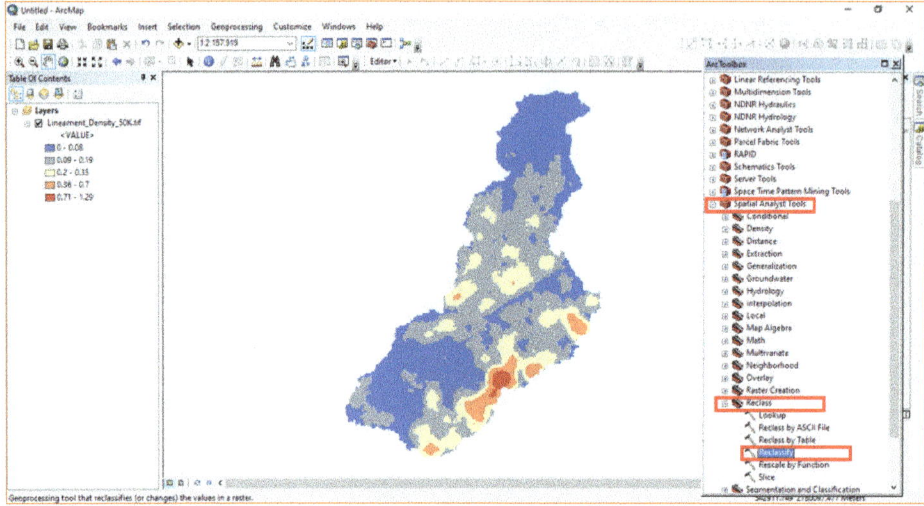

Add 'Lineament_Density_50K.tif' file in **Input raster** option, select value in **Reclass Field** option, and give output Lineament_Density.tif in **Output raster** option, and click on **OK** option.

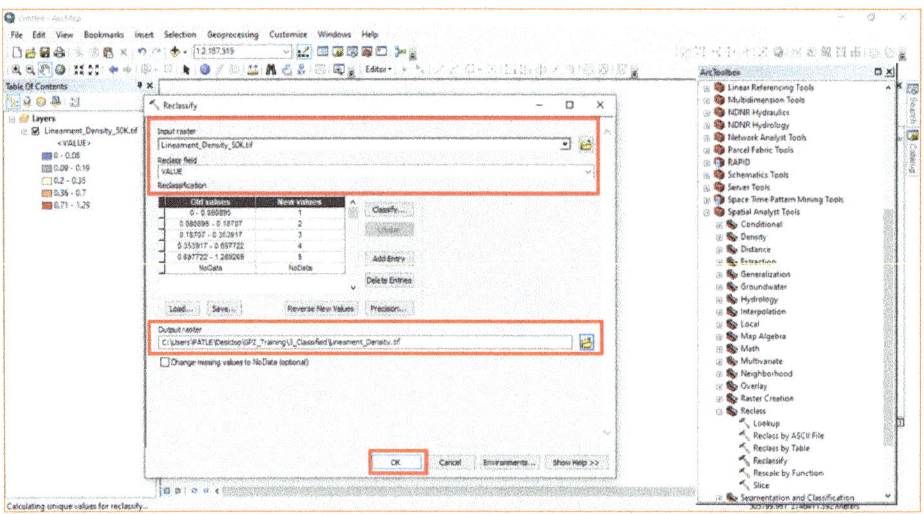

Lineament Density has been reclassified.

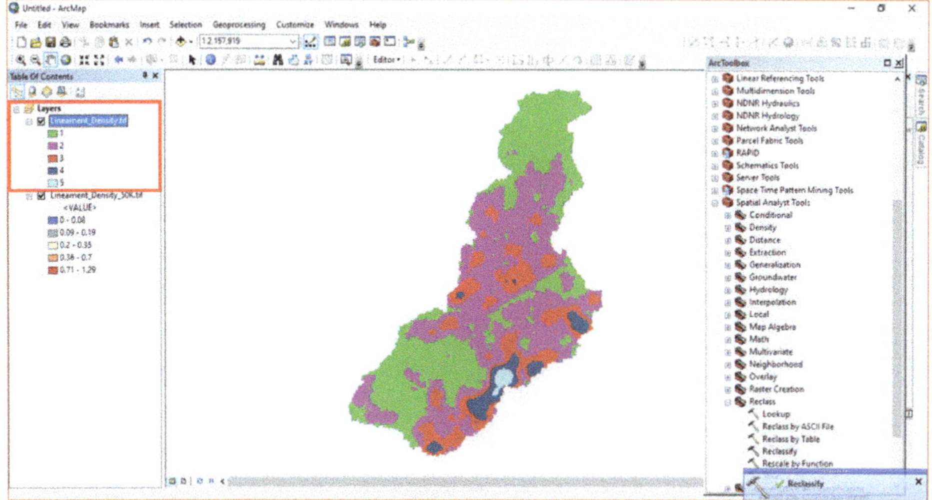

Right click on reclassified layer and click on **Open Attribute Table** option, and click on **Table Option.**

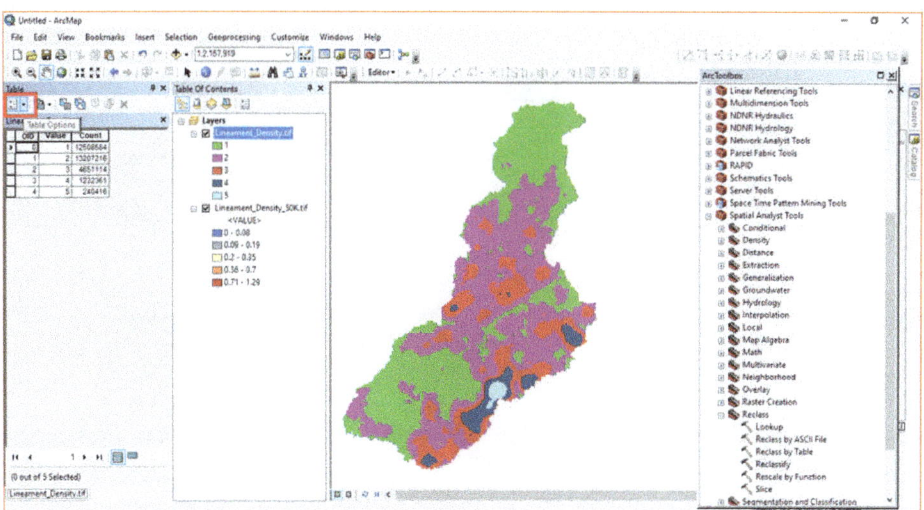

Click on **Add Field** option, give **field name** is LD and type select as **text**, and hit the OK button.

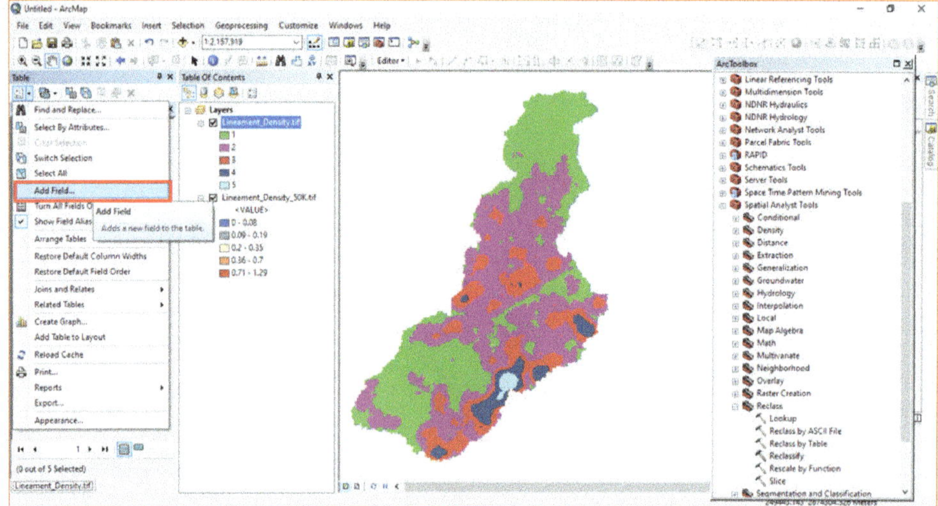

Again, right click on reclassified layer and click on **Open Attribute Table** option. Then, go to **Editor** menu and click on **Start Editing** option.

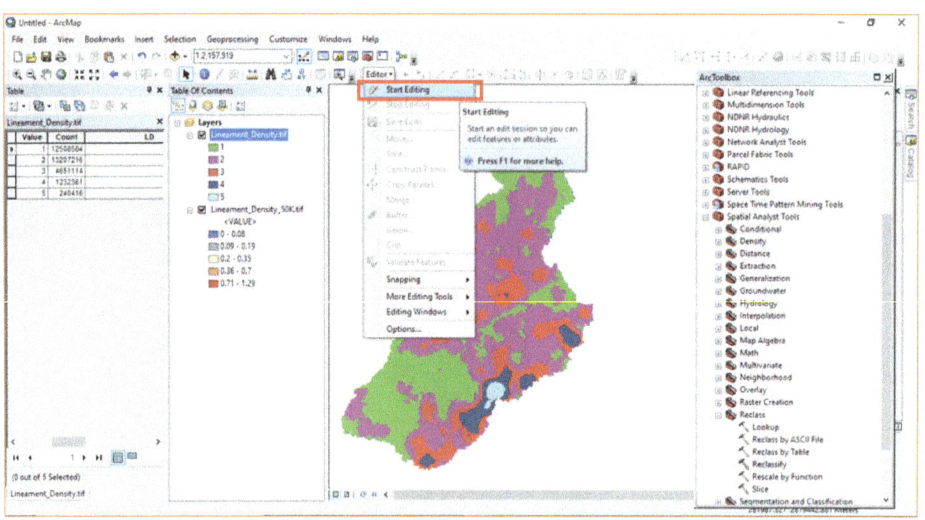

Go to **LD** field, fill the Class/ range of Lineament Density, and click on **Save Edits** option.

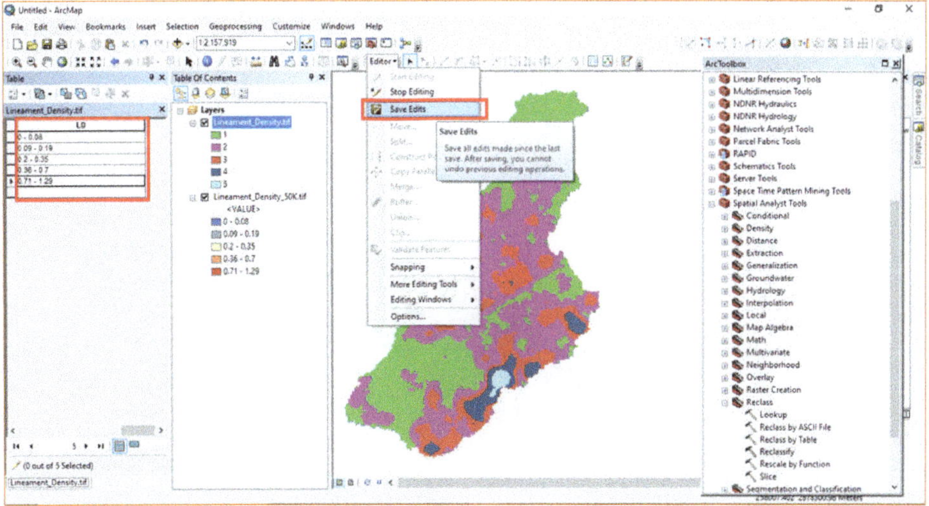

Click on **Stop Editing** option.

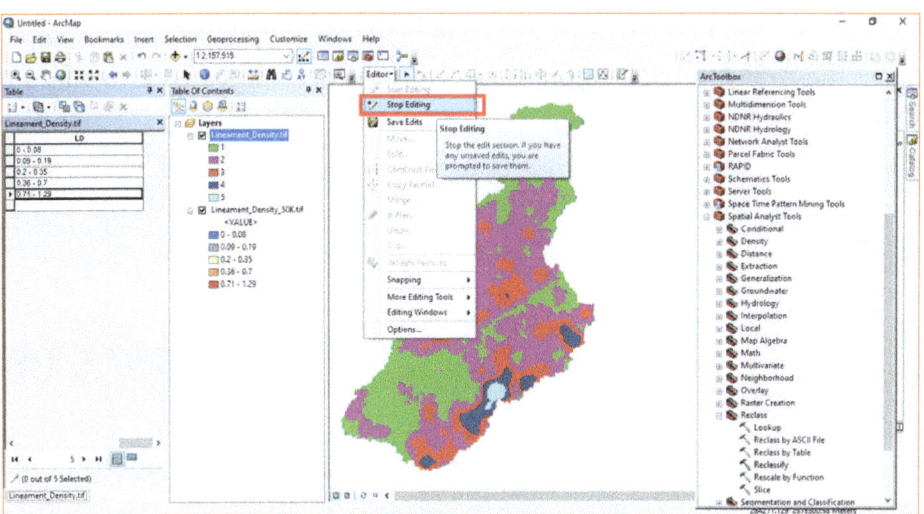

The Lineament Density is properly reclassified.

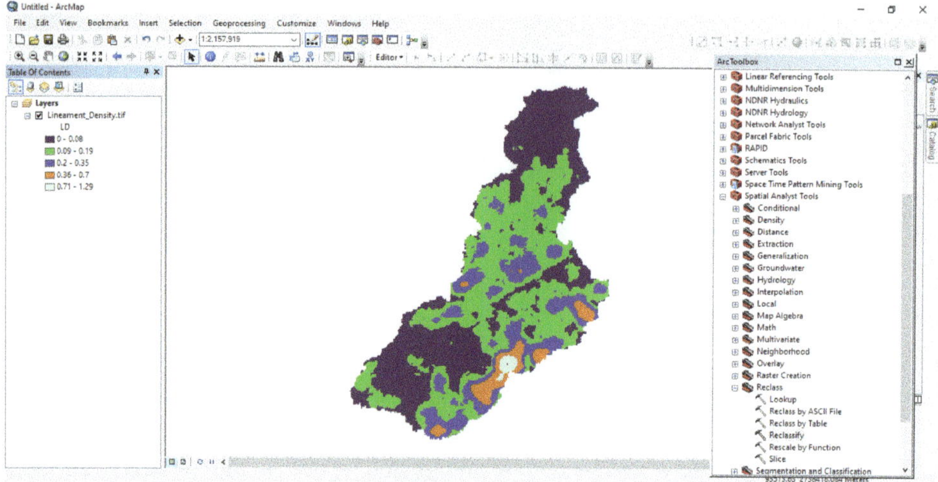

LAND USE/ LAND COVER

In this study, land use/ land cover map of Ken Basin is developed through supervised classification method using LANDSAT imagery.

Open the **ArcMap** application, click on **Add** icon, select **LULC.tif** file, and then click on the **Add** option.

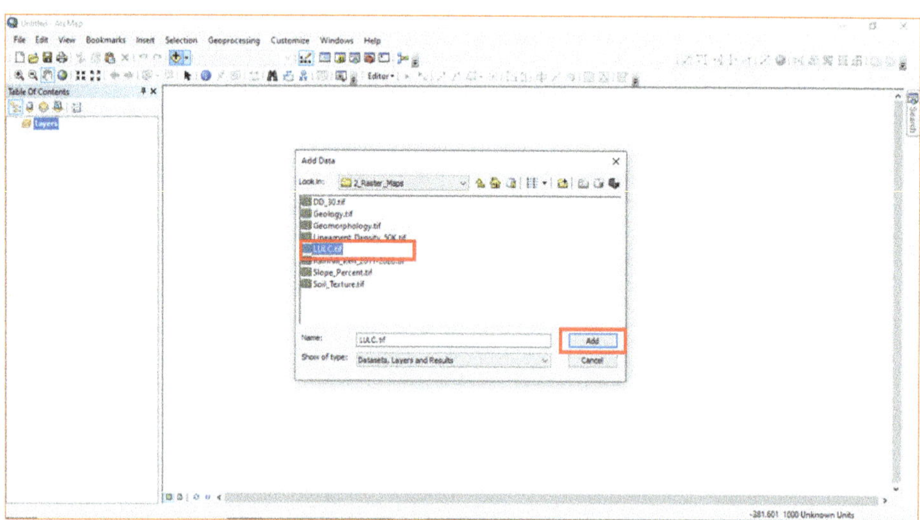

This raster map is already classified into six major classes.

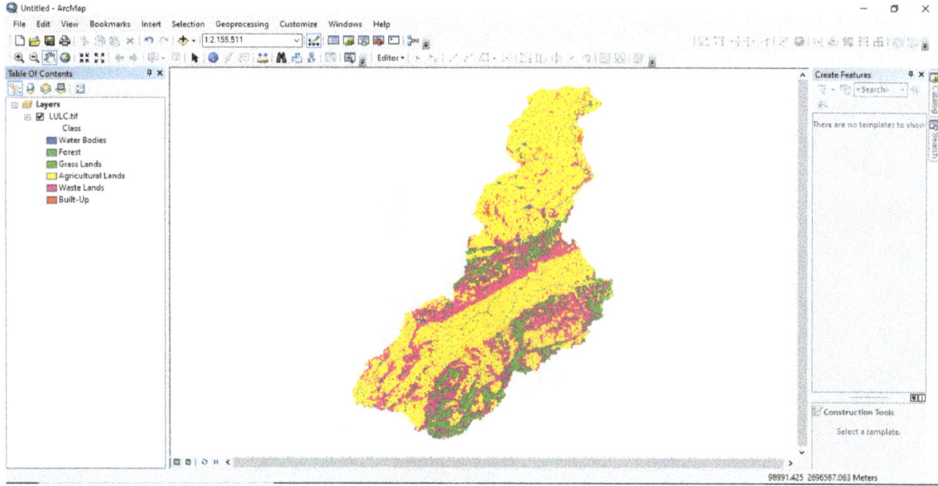

SOIL

Open the **ArcMap** application, click on **Add** icon, select 'Soil_Texture.tif' file, and then click on **Add** option.

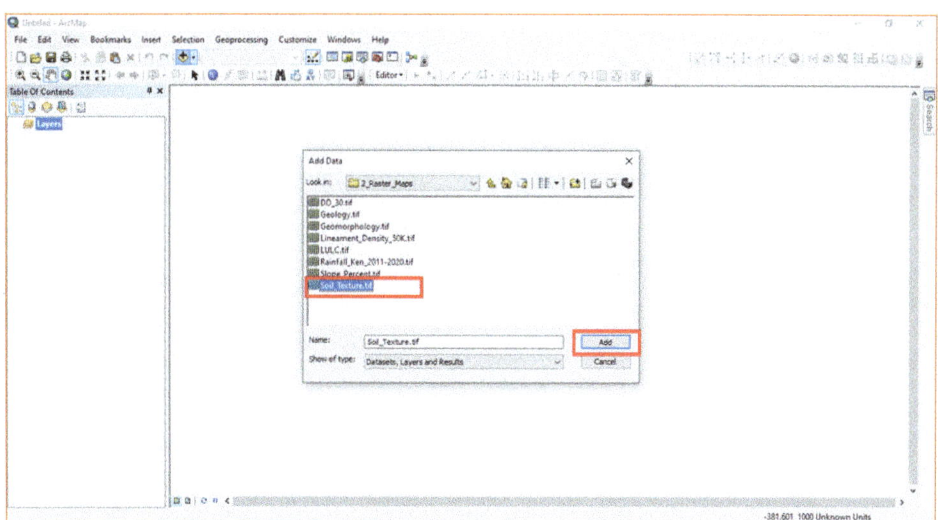

This raster map is already classified into four major textural classes.

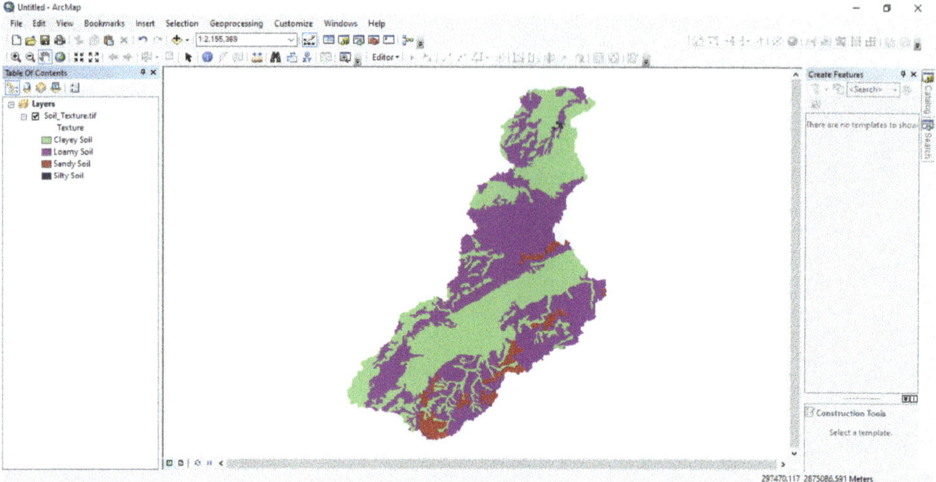

SLOPE

Browse the 'Raster_Maps' folder, select on the **Slope_Percent.tif** file, and then click on **Add** option.

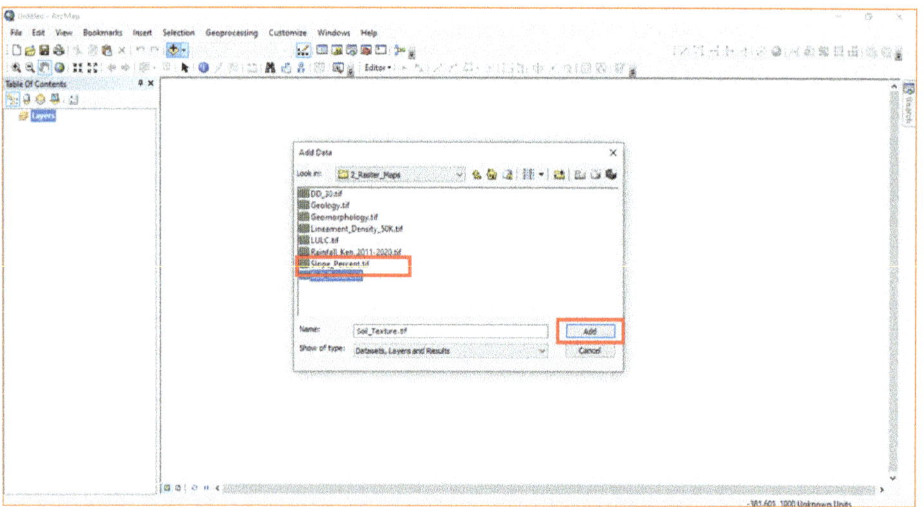

The slope map is available in dynamic composite map form. It needs to be reclassified.

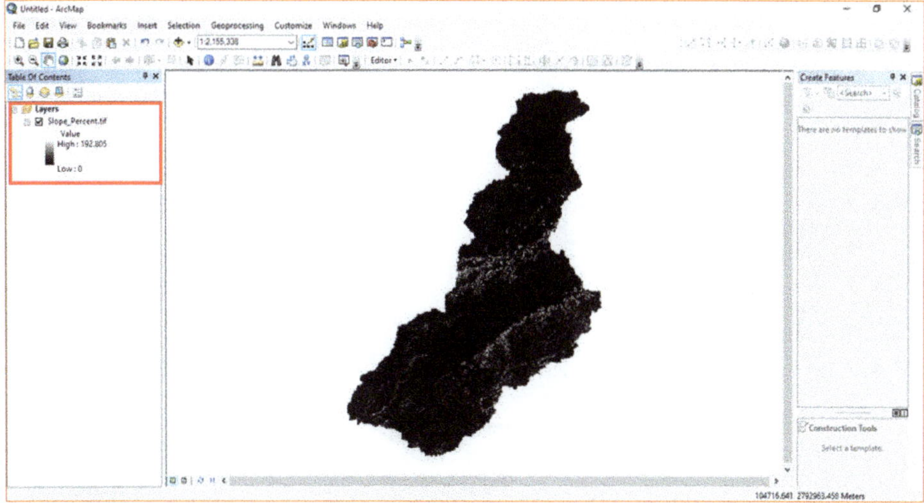

Right click on **Slope_Percent.tif** file showing in Layer panel. Click on **Properties** option.

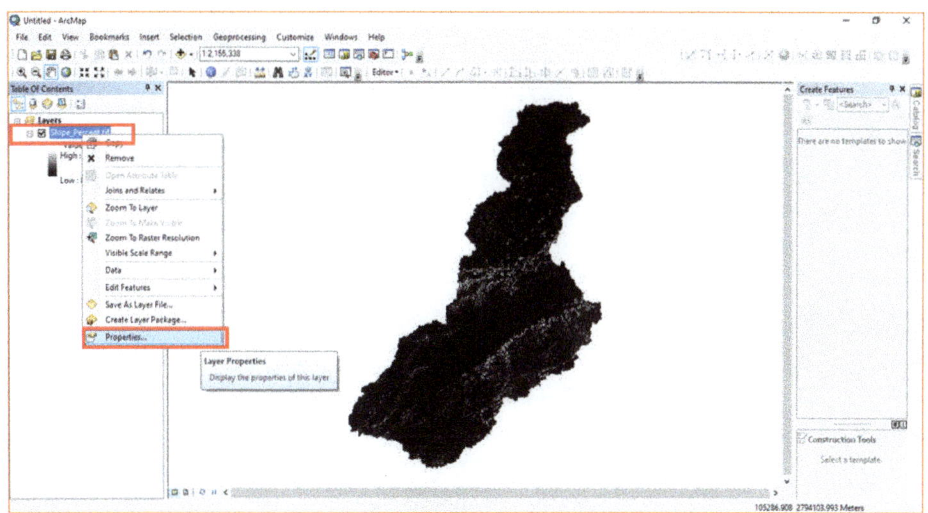

Click on **Symbology** option, select the **Classified** option. and then click on **Classify** option.

Select **number of classes** in class option.

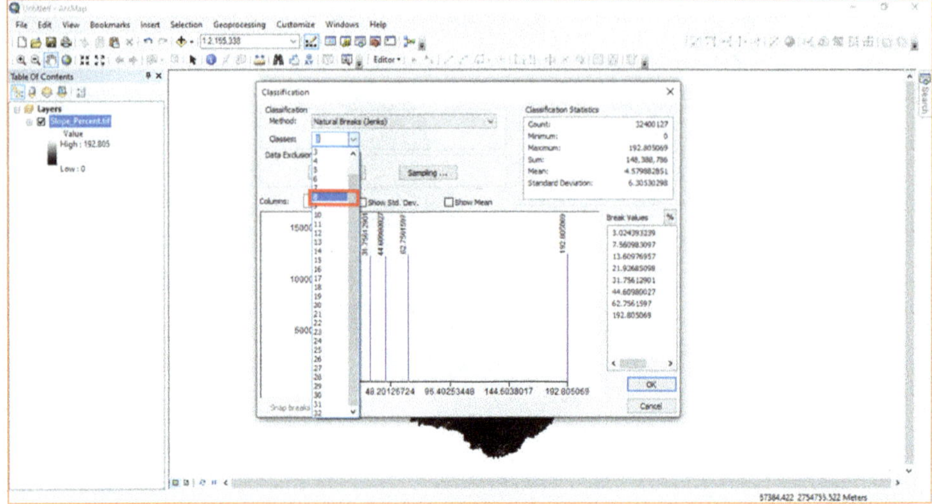

Select **Manual** option in classification method.

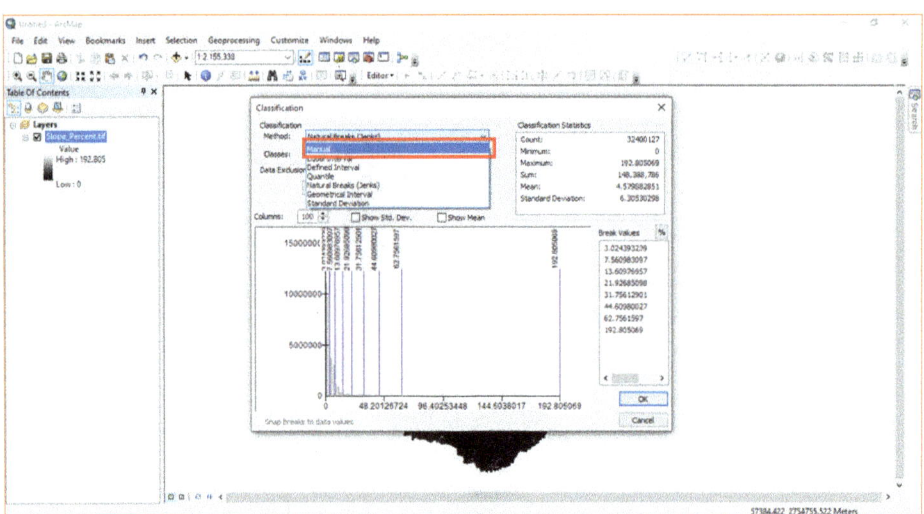

Go to **Breaks Values %** panel, give the break values 0 to > 25% according to land capability classification system, and click on **OK** option.

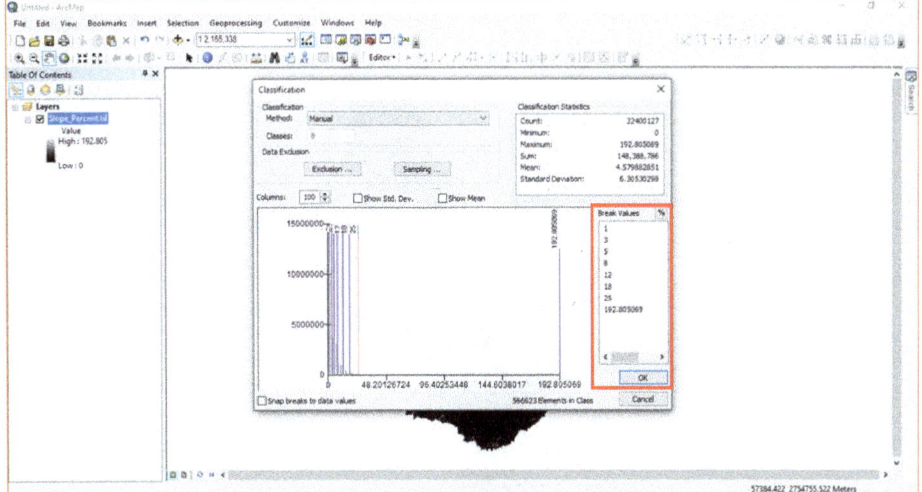

Change the color from **Color Ramp**, click on **Apply** option, and hit the **OK** option.

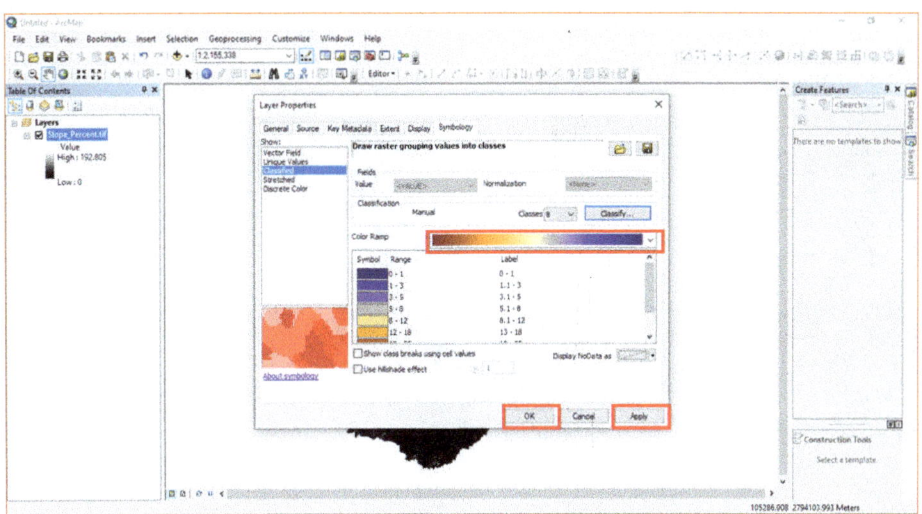

Slope map of Ken Basin has been classified in Eight Classes.

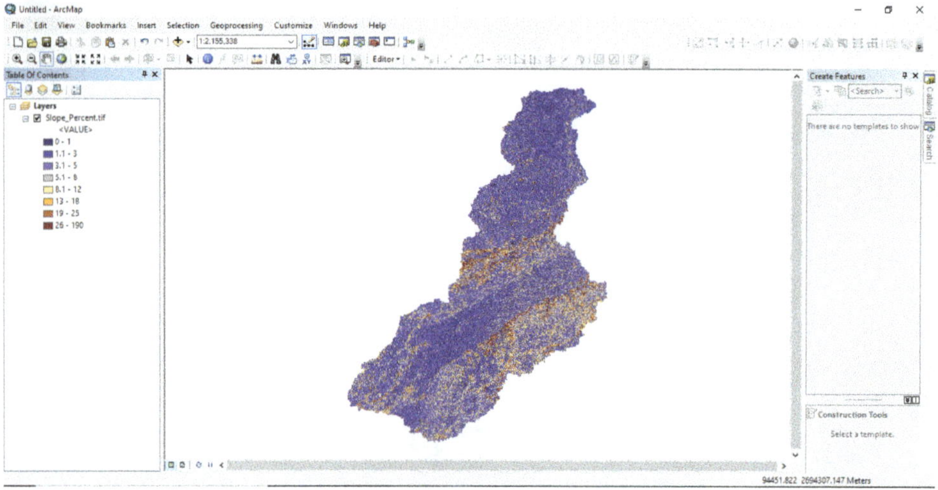

Go to:

ArcTool Box > Spatial Analyst Tools > Reclass > Reclassify. Double-Click on Reclassify tool.

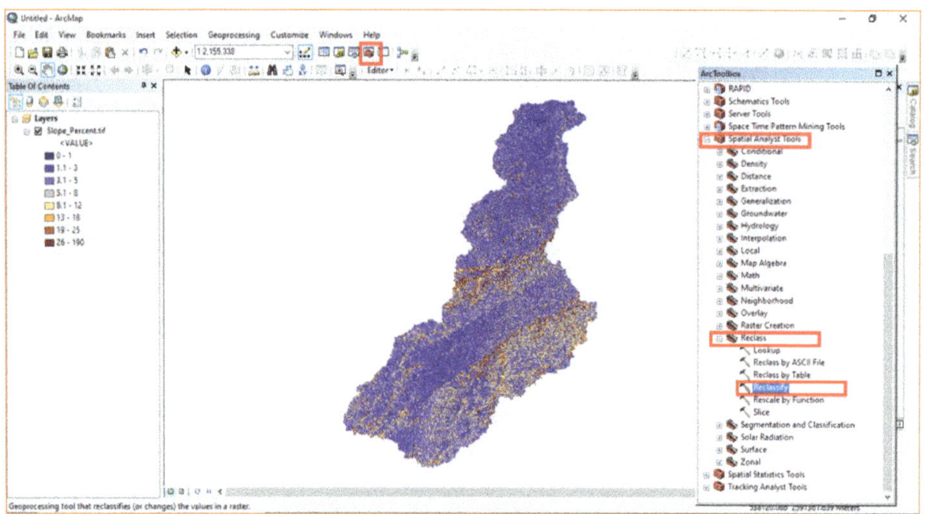

Add 'Slope_Percent.tif' file in **Input raster** option, select value in **Reclass Field** option, and give output Slope_Percent.tif in **Output raster** option, and click on **OK** option.

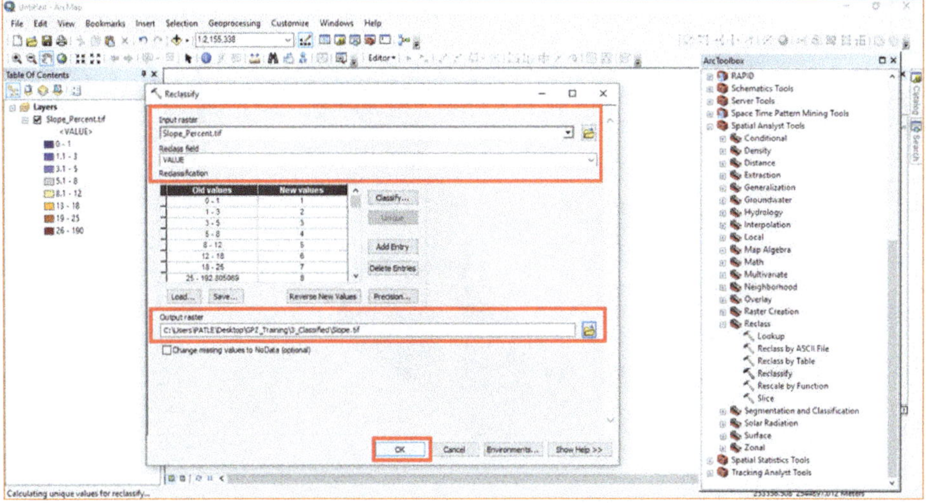

Slope map has been reclassified.

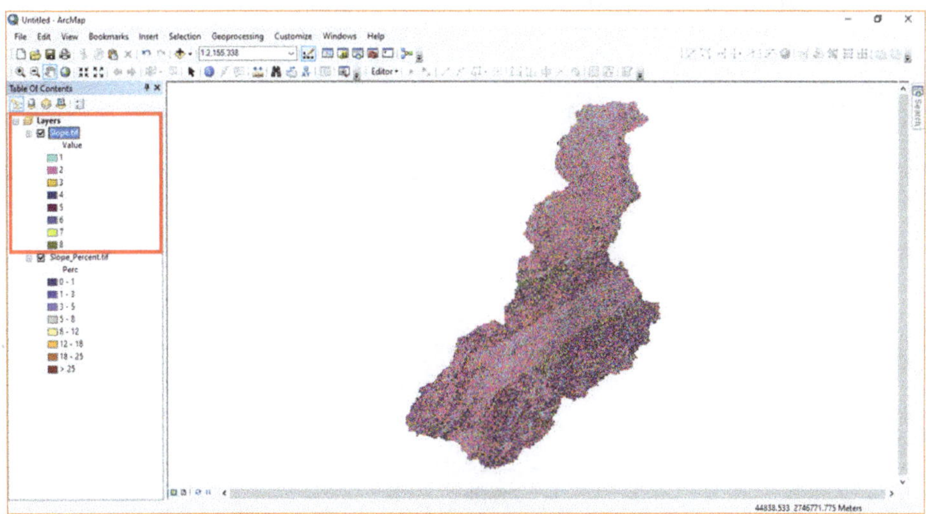

Right click on reclassified layer and click on **Open Attribute Table** option, and click on **Table Option**.

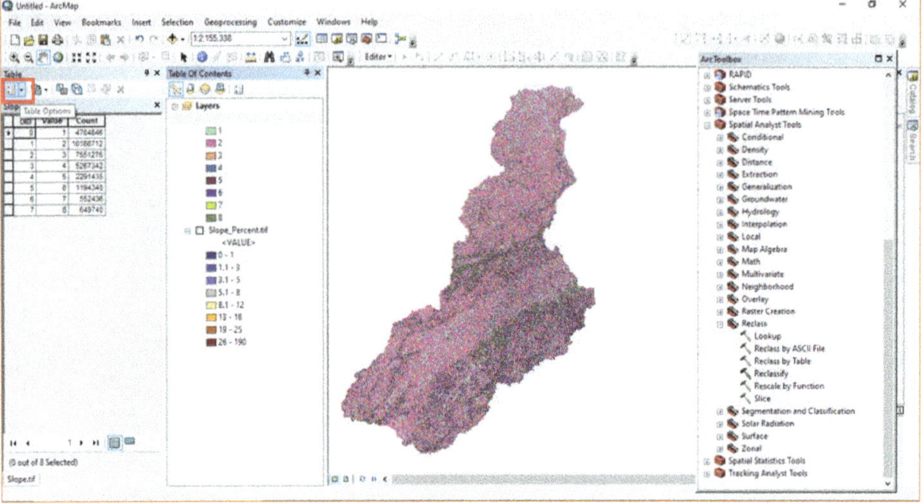

Click on **Add Field** option, give **field name** is Slope_Perc and type select as **text**, and hit the **OK** button.

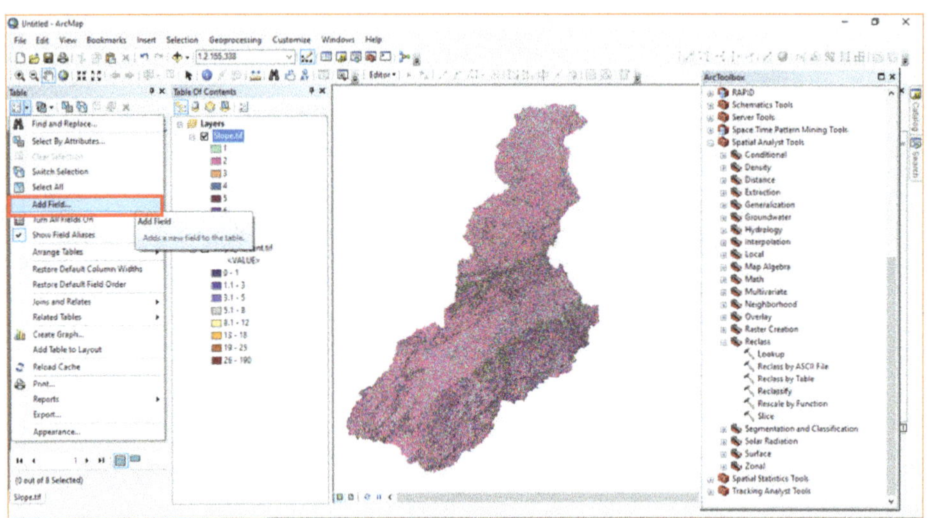

Again, right click on reclassified layer and click on **Open Attribute Table** option. Then, go to **Editor** menu and click on **Start Editing** option.

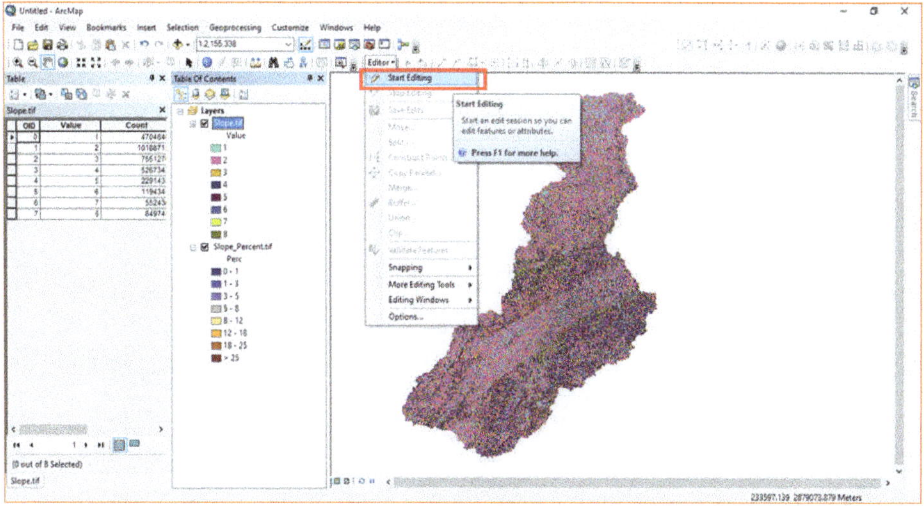

Go to **Slope_Perc** field, fill the Class/ range of Slope map, and click on **Save Edits** option.

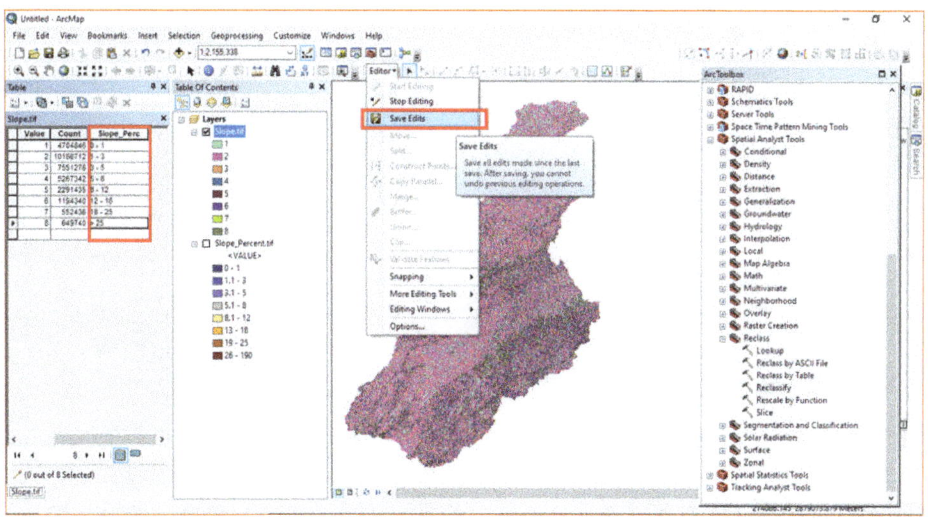

Click on **Stop Editing** option.

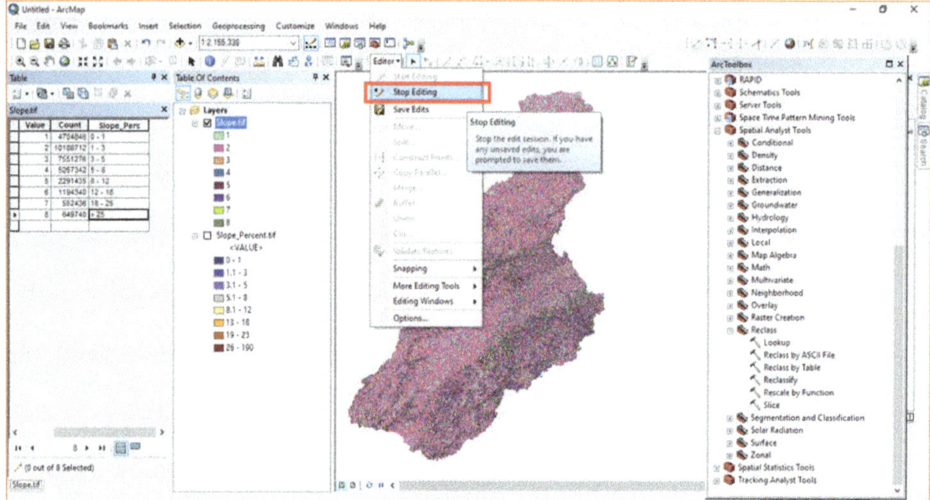

The Slope is properly reclassified.

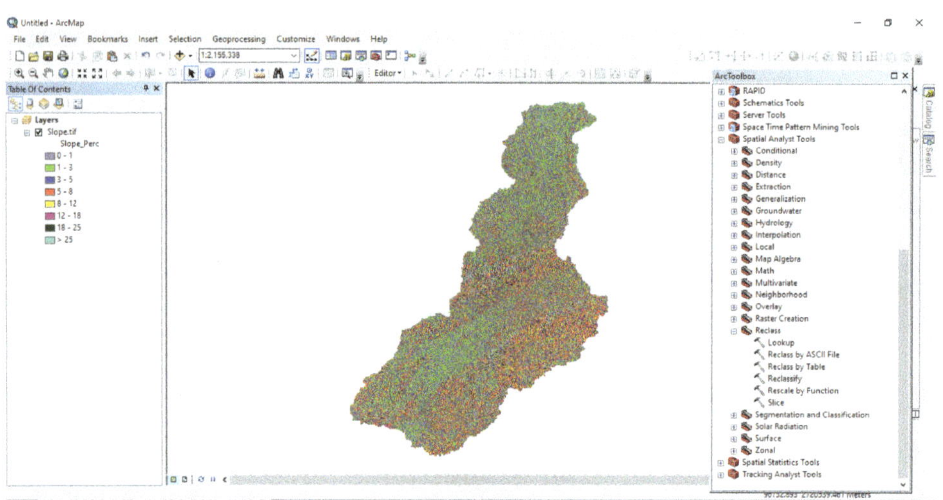

DRAINAGE DENSITY

Browse the 'Raster_Maps' folder, select on the **Drainage_Density.tif** file and click on **Add** option.

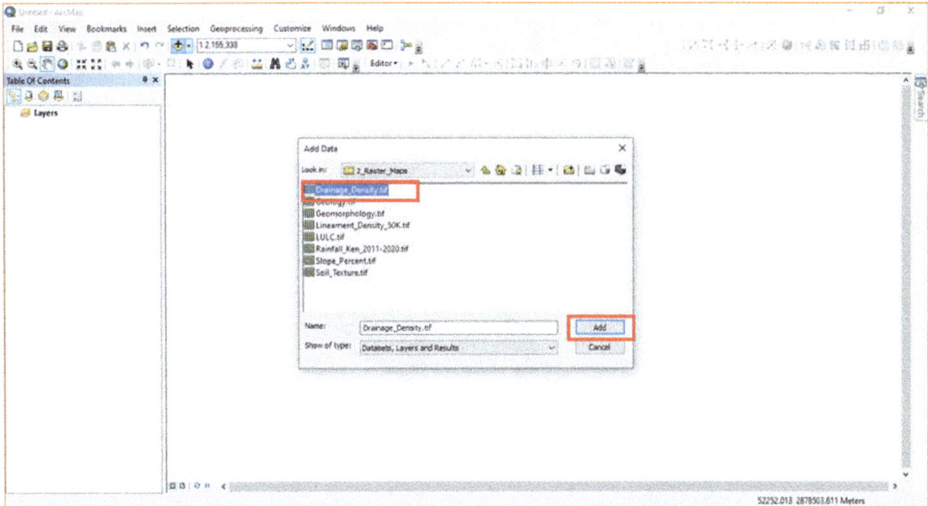

Drainage Density map is available in the form of dynamic composite map. It needs to be reclassified.

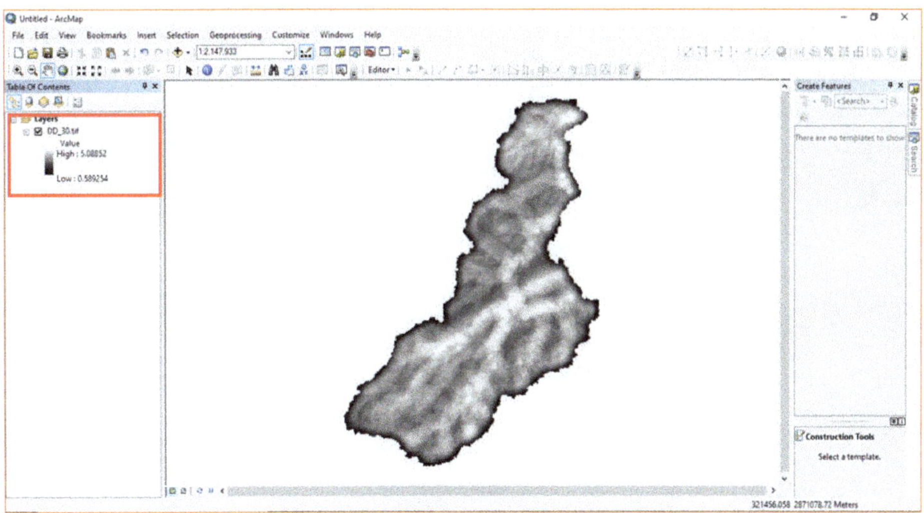

Right click on the **Drainage_Density.tif** file showing in Layer panel. Click on **Properties** option

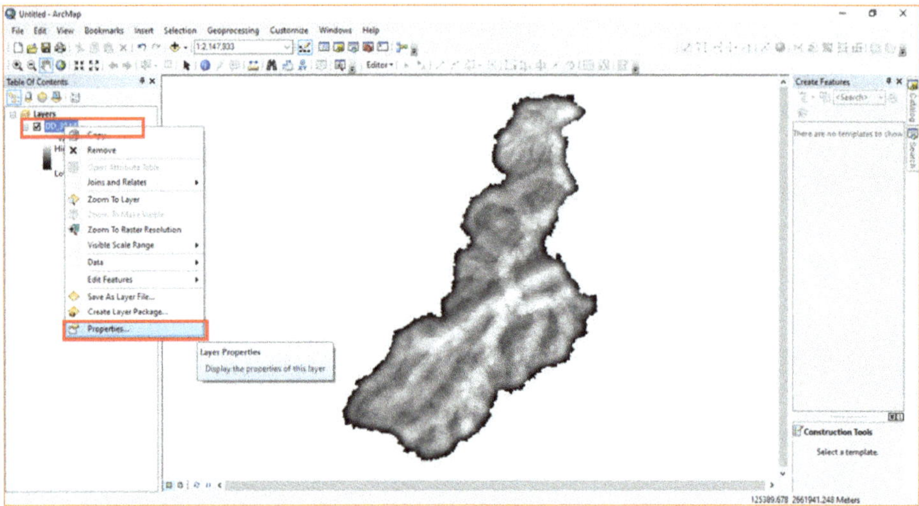

Click on **Symbology** option, select the **Classified** option, give the number of **Classes**, change the color from **Color Ramp**, click on **Apply** option, and hit the **OK** option.

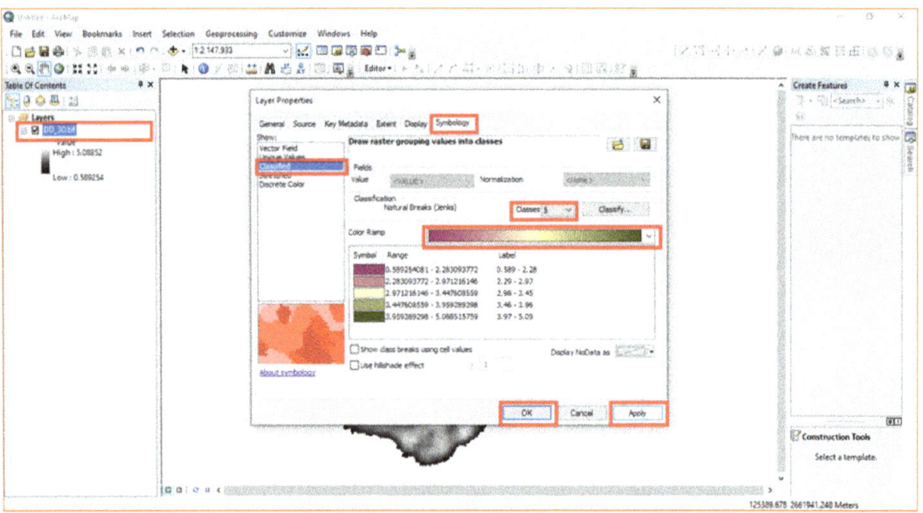

The drainage density of Ken basin has been classified into 5 classes using the **Natural Break (Jenks)** classification method. It needs to be **reclassified.**

Classification method may be changed according to the purpose of the study.

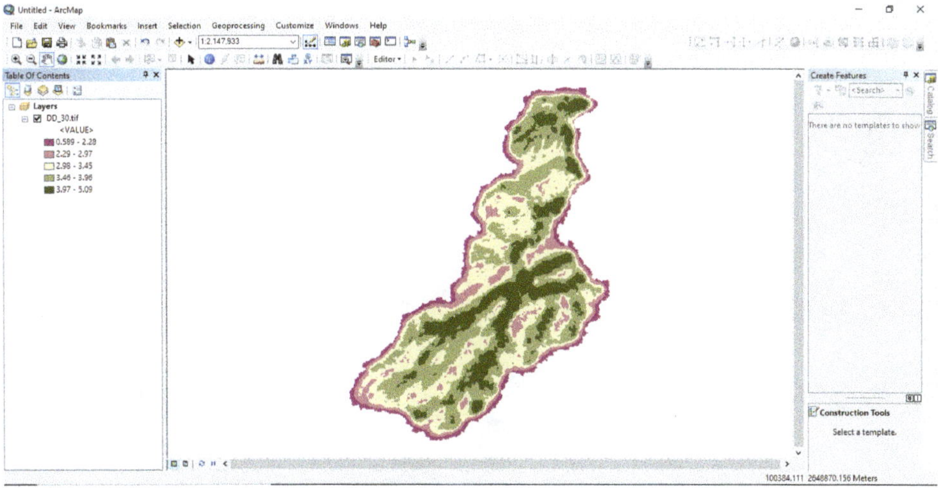

Click on **ArcTool Box** icon.

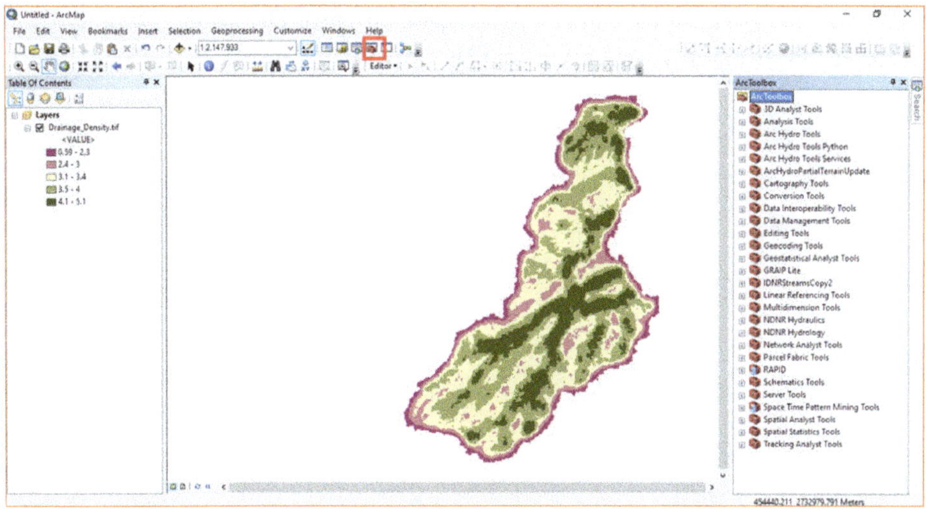

Go to:

ArcTool Box > Spatial Analyst Tools > Reclass > Reclassify. Double-Click on Reclassify tool.

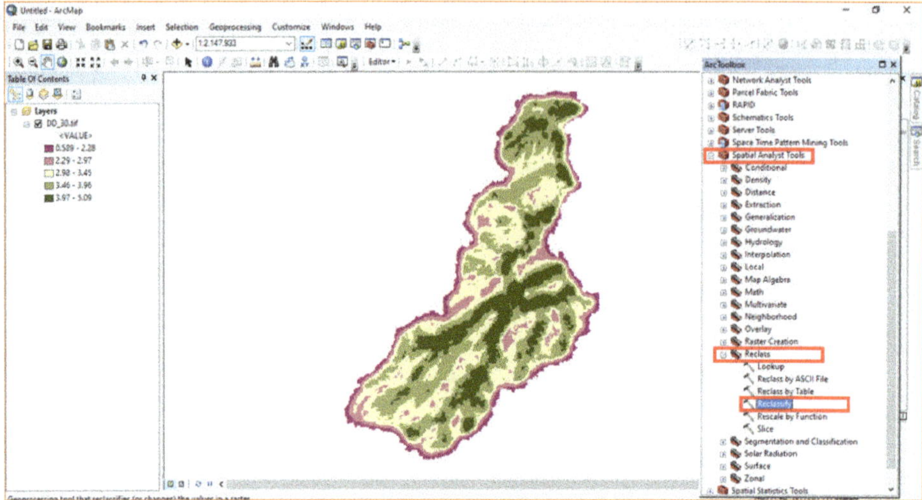

Add Drainage Density layer 'DD_30.tif' file in **Input raster** option, select value in **Reclass Field** option, and give output Drainage_Density.tif in **Output raster** option, and click on **OK** option.

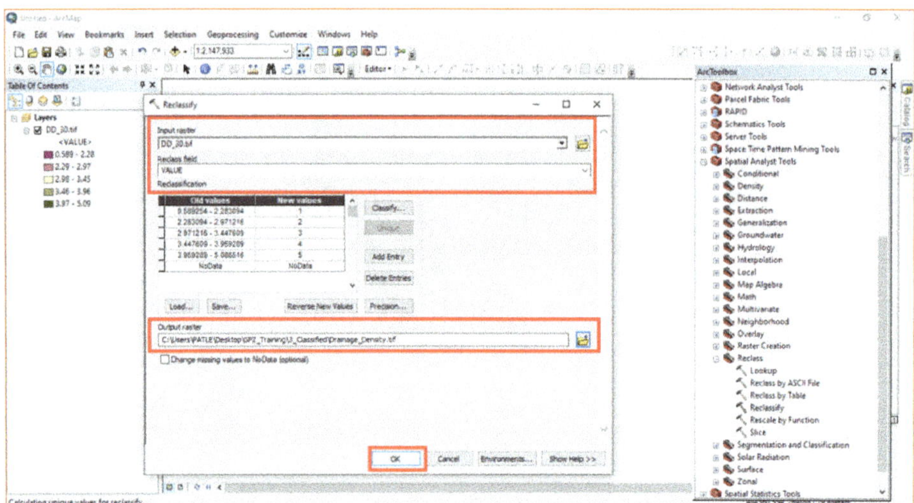

Drainage Density has been reclassified.

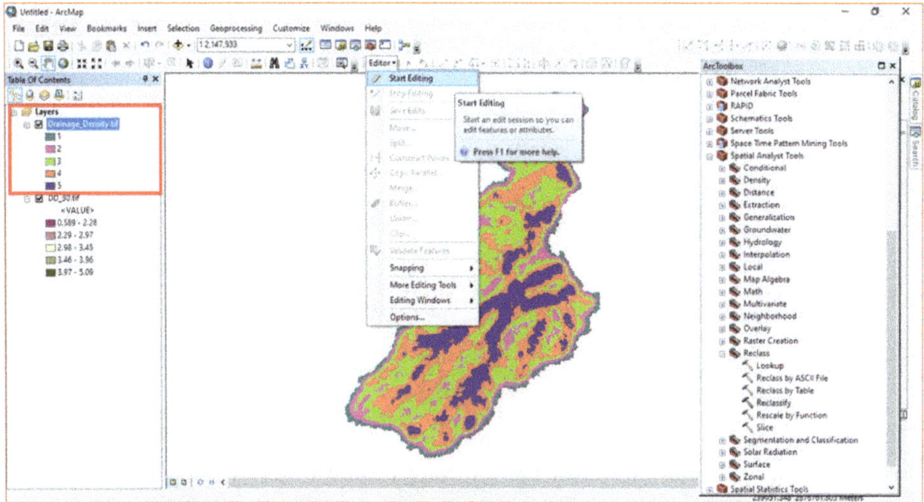

Right click on reclassified layer and click on **Open Attribute Table** option, and click on **Table Option**.

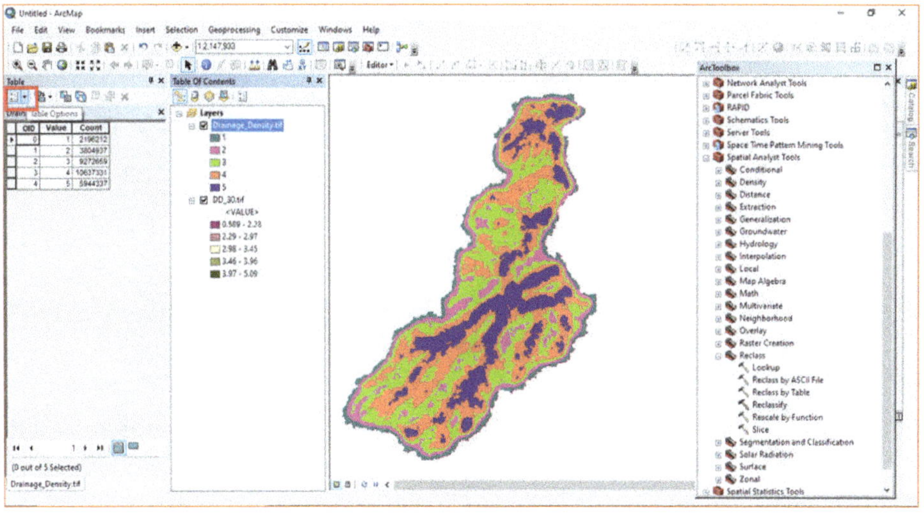

Click on **Add Field** option, give **field name** is DD and type select as **text**, and hit the OK button.

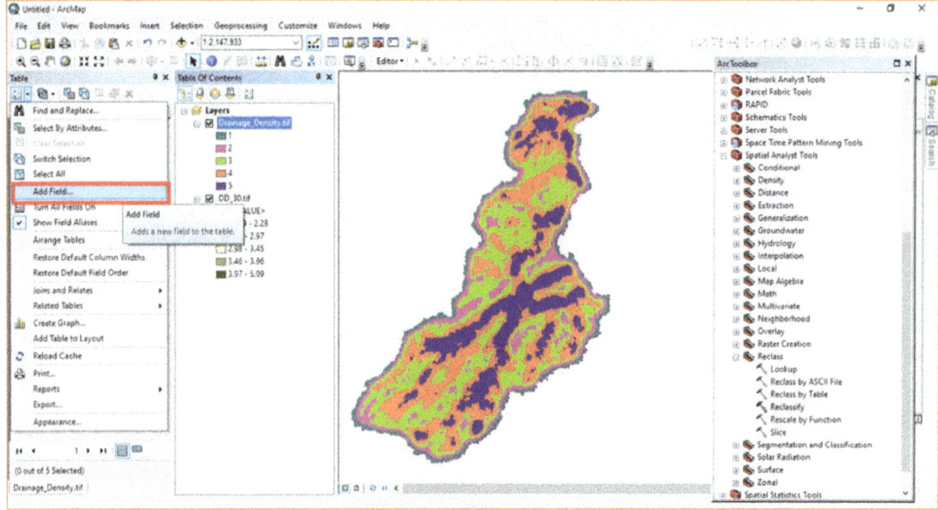

Again, right click on reclassified layer and click on **Open Attribute Table** option. Then, go to **Editor** menu and click on **Start Editing** option.

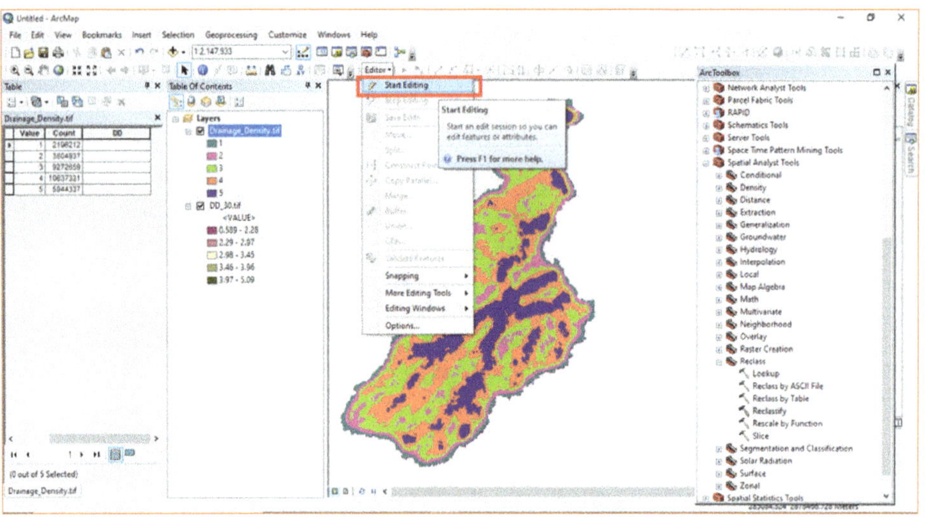

Go to **DD** field, fill the Class/ range of Drainage Density, and click on **Save Edits** option.

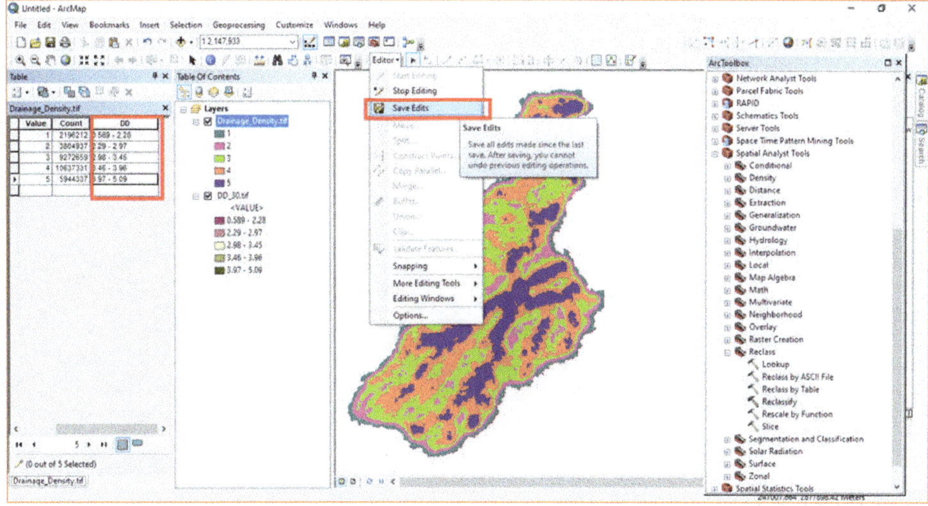

Click on **Stop Editing** option.

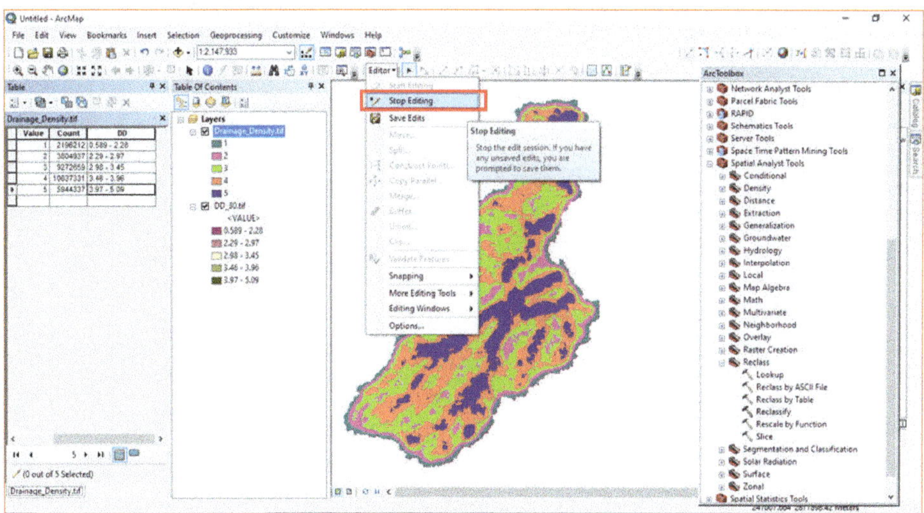

The Drainage Density is properly reclassified.

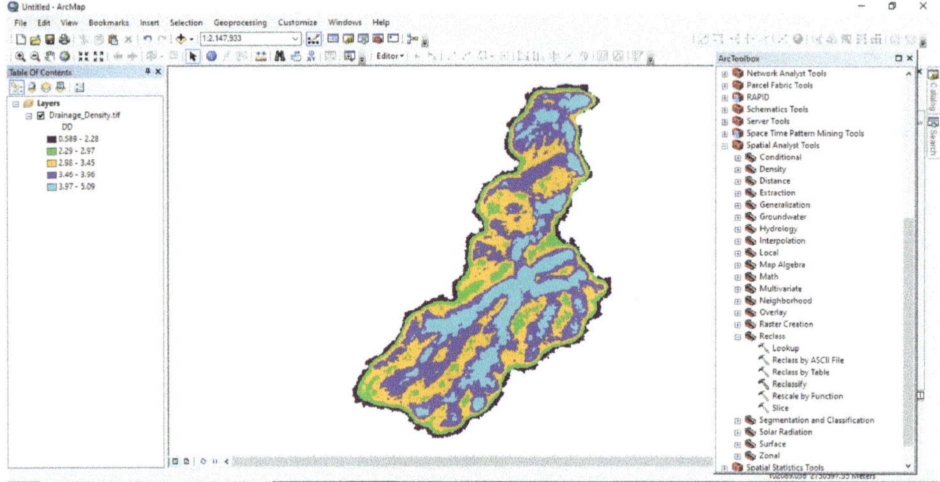

RAINFALL

Browse the 'Raster_Maps' folder, select on the **Rainfall_Ken_2011-20.tif** file and click on **Add** option.

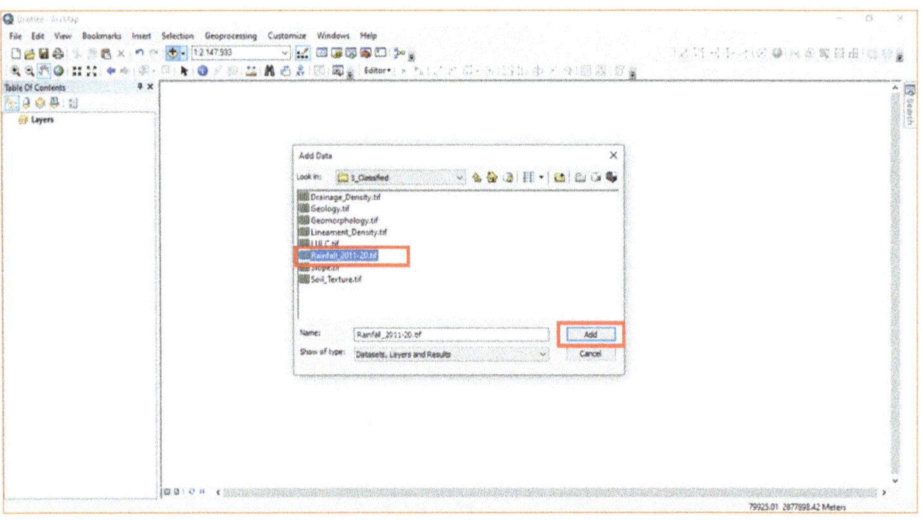

Rainfall map is available in the form of dynamic composite map. It needs to be reclassified.

Right click on **Rainfall_Ken_2011-20.tif** file showing in Layer panel. Click on **Properties** option

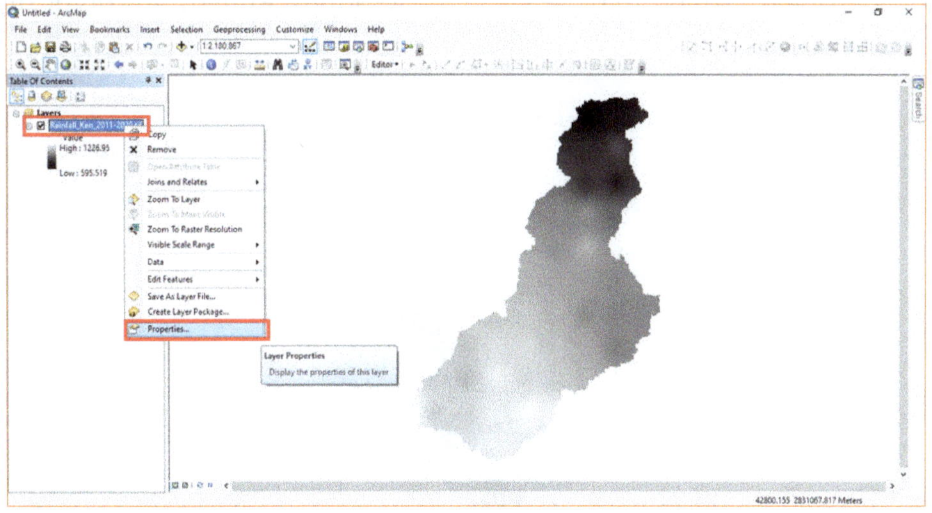

Click on **Symbology** option, select the **Classified** option, give the number of **Classes**, change the color from **Color Ramp**, click on **Apply** option, and hit the **OK** option.

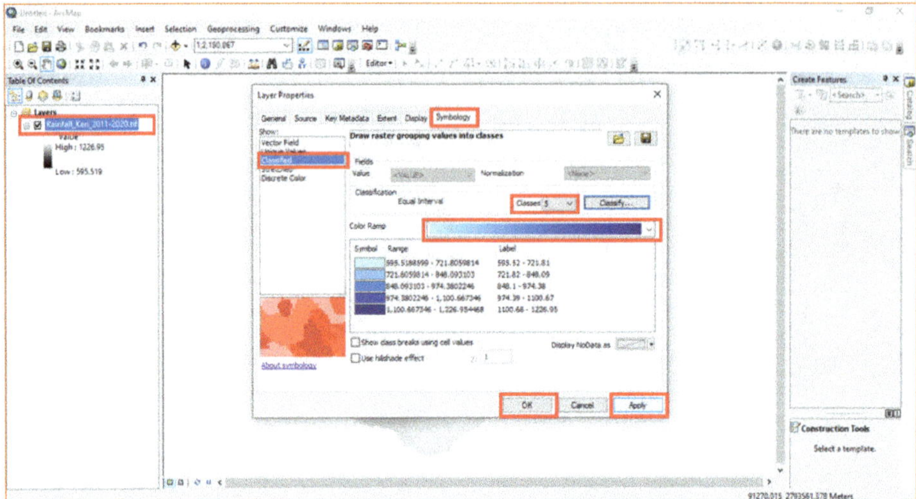

The rainfall map of Ken basin has been classified into 5 classes using the **Equal Intervals** classification method. It needs to be **reclassified**.

Classification method may be changed according to the purpose of the study.

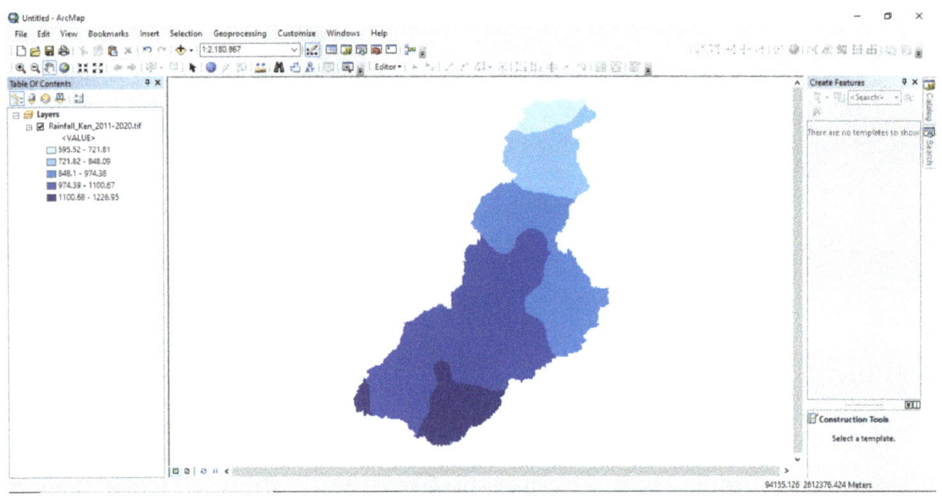

Click on **ArcTool Box** icon.

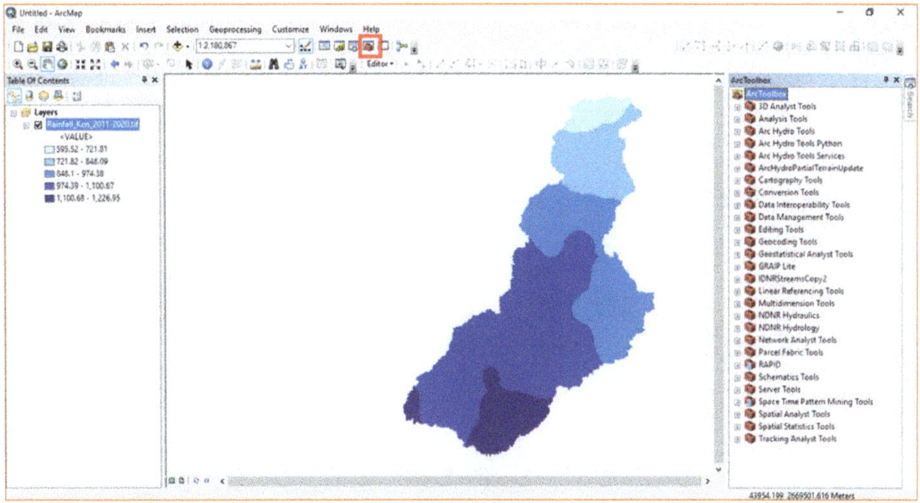

Go to:

ArcTool Box > Spatial Analyst Tools > Reclass > Reclassify. Double-Click on Reclassify tool.

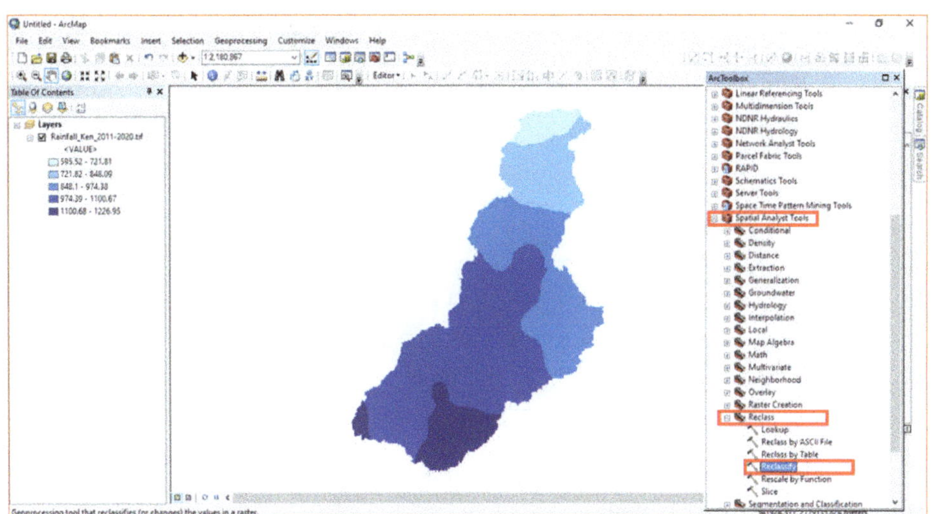

Add 'Rainfall_Ken_2011-20.tif' file in **Input raster** option, select value in **Reclass Field** option, and give output 'Rainfall_2011-20.tif' in **Output raster** option, and click on **OK** option.

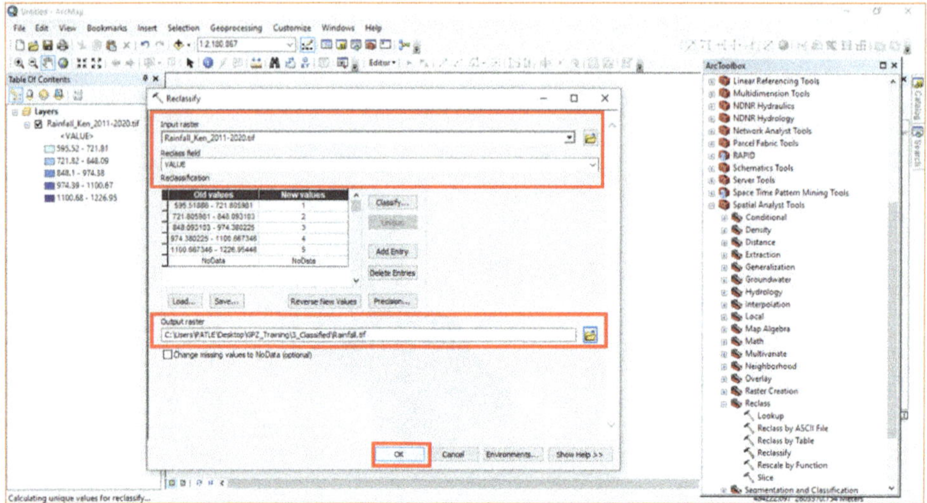

Rainfall map has been reclassified.

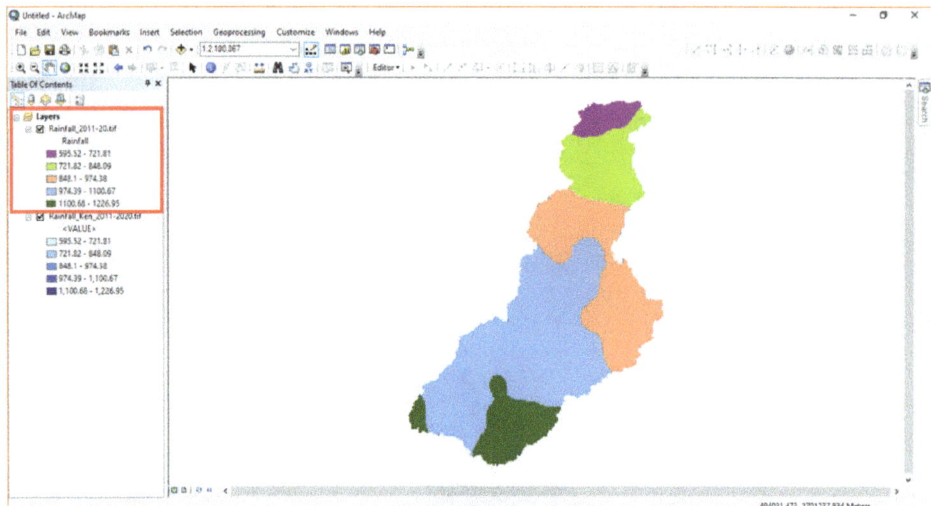

Right click on reclassified layer and click on **Open Attribute Table** option, and click on **Table Option**.

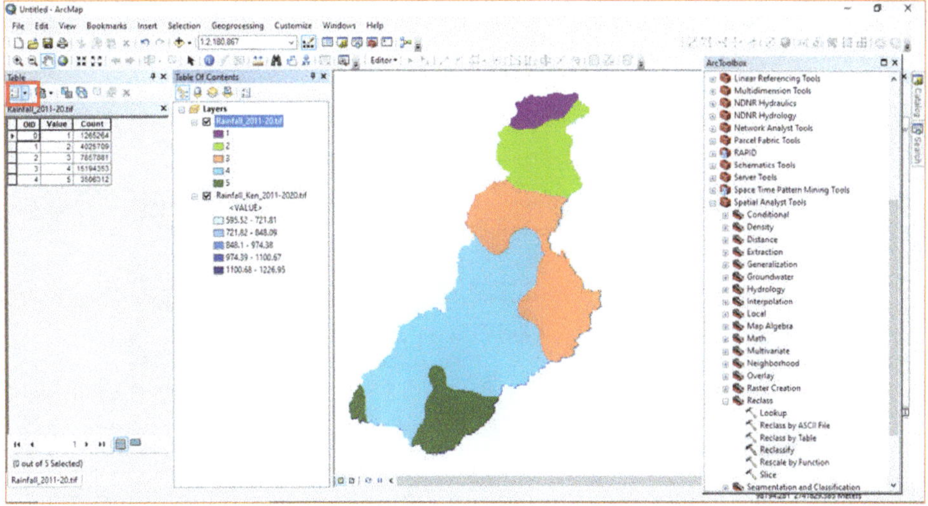

Click on **Add Field** option, give **field name** is Rainfall and type select as **text**, and hit the OK button.

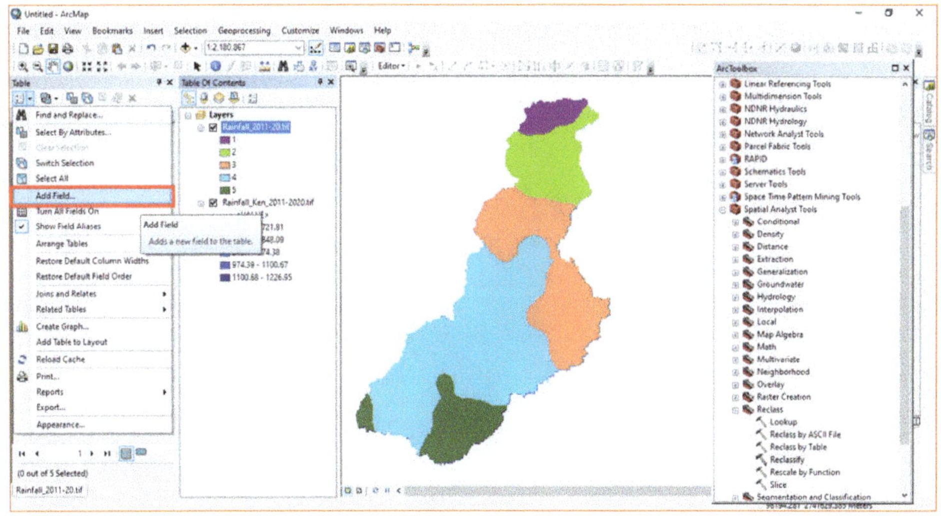

Again, right click on reclassified layer and click on **Open Attribute Table** option. Then, go to **Editor** menu and click on **Start Editing** option.

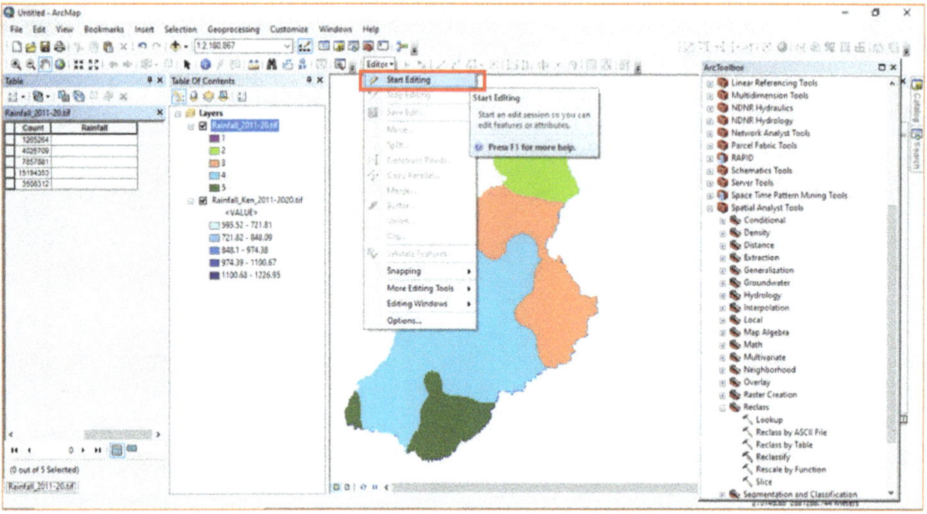

Go to **Rainfall** field, fill the Class/ range of Rainfall in mm, and click on **Save Edits** option.

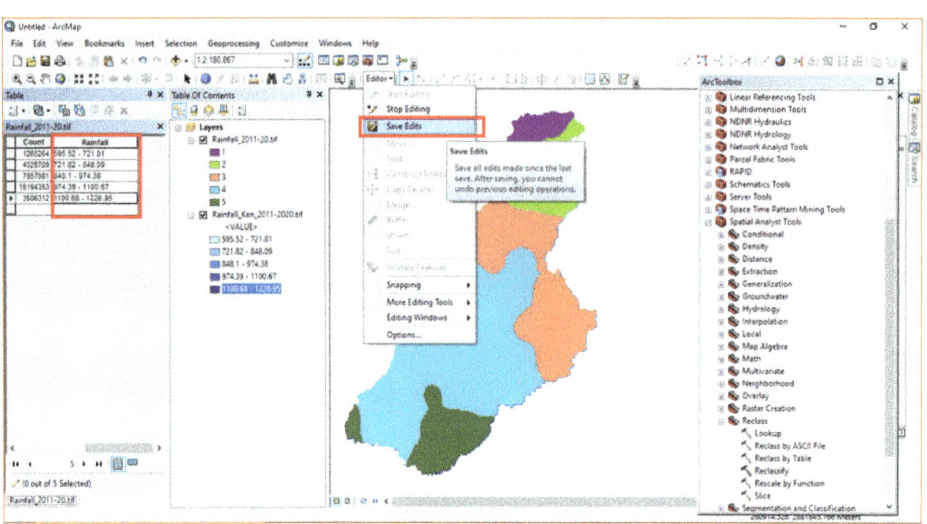

Click on **Stop Editing** option.

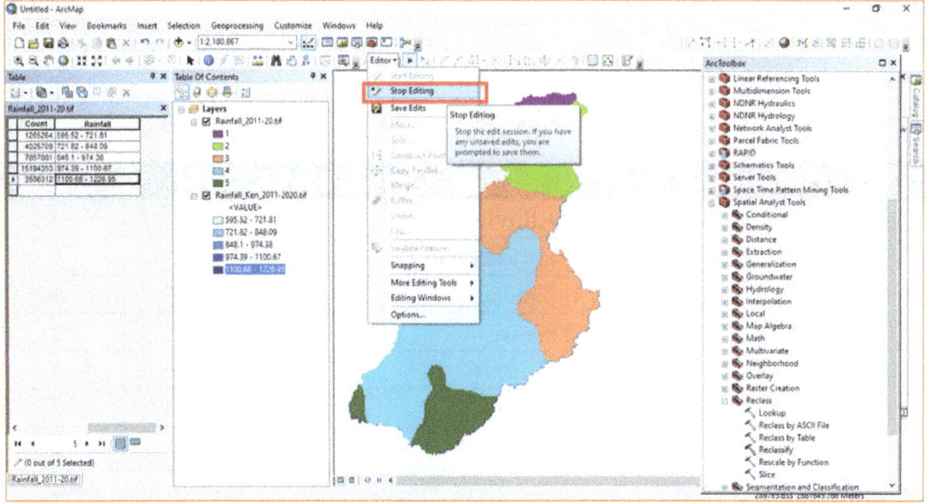

The Rainfall is properly reclassified.

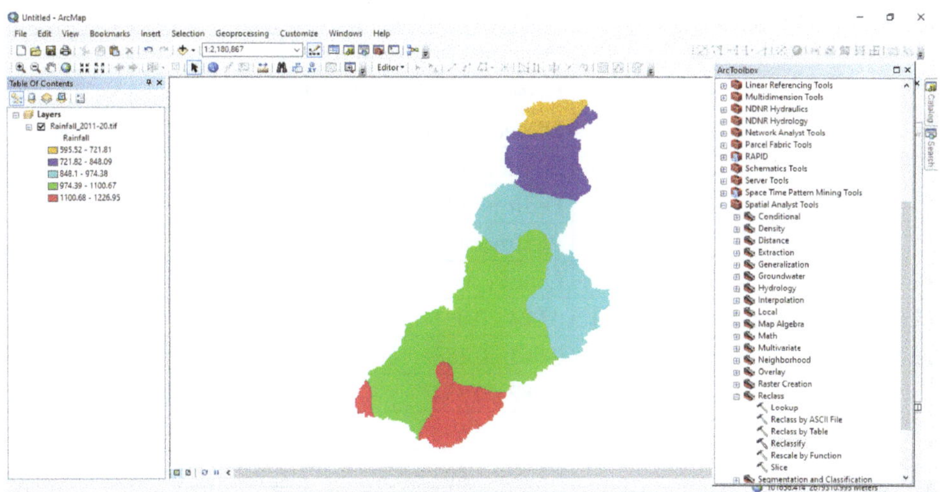

RANKING OF THEMATIC MAPS

Rank assignment for different thematic layers was executed based on the relative importance of groundwater. Rankings of thematic factors were assigned from 5 to 1 in the order of most influencing class/ category to least influencing class/category. The rankings and of different thematic factors (namely geology, geomorphology, lineament density, land use/ land cover, soil texture, slope, drainage density, rainfall and water level fluctuation) were enlisted in the following Table 1.

Table 1: Rankings for different layers

Factors	Classes/ Ranges	Rank
Geology	Alluvium	5
	Bundelkhand Granitoid Complex	1
	Bijawar	3
	Semri	2
	Kaimur	5
	Rewa	3
	Malwa	2
	Bhander	4
	Lameta	1
	Laterite	4
Geomorphology	Alluvial Plain	4
	Dam and Reservoir	5
	Dissected Hills and Valleys	1
	Dissected Plateau	2
	Flood Plain	5
	Pediment Pediplain Complex	3
	Piedmont Slope	2
	Quarry and Mine Dump	1
Lineament Density (km/ km^2)	0.024 - 0.081	1
	0.081 - 0.187	2
	0.187 - 0.354	3
	0.354 - 0.698	4
	0.698 - 1.289	5
Land Use/ Land Cover	Water Bodies	5
	Forest	3
	Grazing lands	3
	Agricultural lands	4
	Waste lands	2
	Built-up	1

Soil Texture	Clayey Soil	1
	Loamy Soil	3
	Sandy Soil	4
	Silty Soil	2
Slope (%)	0 - 1	5
	1 - 3	4
	3 - 5	3
	5 - 8	2
	8 - 12	1
	12 - 18	1
	18 - 25	1
	> 25	1
Drainage Density (km/ km²)	0.589 - 2.283	5
	2.283 - 2.971	4
	2.971 - 3.450	3
	3.450 - 3.960	2
	3.960 - 5.089	1
Rainfall (mm)	< 725	1
	725 - 855	2
	855 - 985	3
	985 - 1115	4
	> 1115	5

The assigned ranks need to be integrated into the raster of all the thematic layers.

GEOLOGY

Open the classified raster layer of Geology in Arc Map application and Click on **ArcTool Box** icon. Then Go to:

ArcTool Box > Spatial Analyst Tools > Reclass > Reclassify. Double-Click on Reclassify tool.

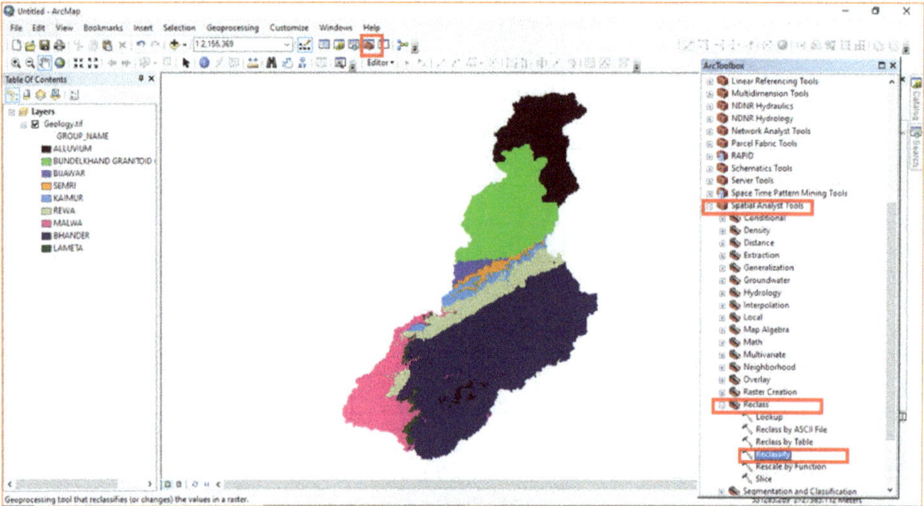

Add 'Geology.tif' file in **Input raster** option, select Group_Name in **Reclass Field** option, assign ranks to each class in **New Values** field according to Table 1, and give output 'Geology_rank.tif' in **Output raster** option, and click on **OK** option.

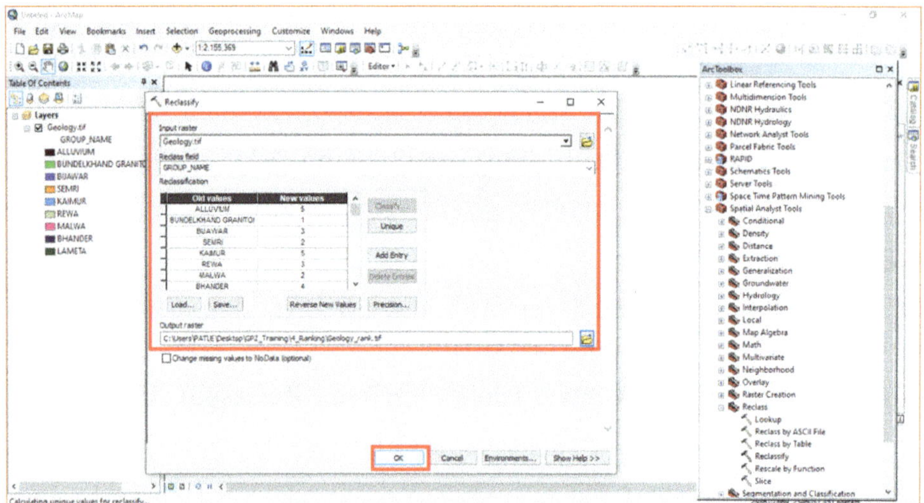

Geology layer has been reclassified according to priority ranks.

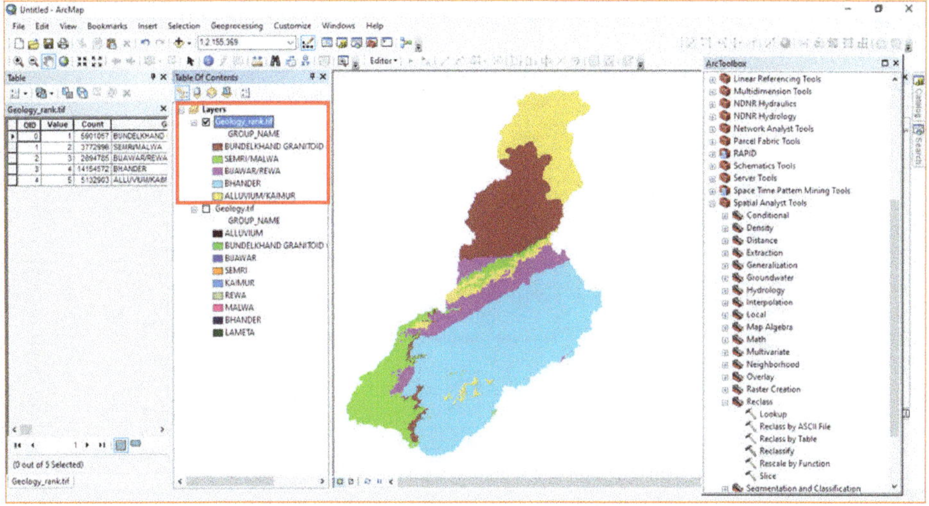

GEOMORPHOLOGY

Open the classified raster layer of Geomorphology in Arc Map application and Click on **ArcTool Box** icon. Then Go to:

ArcTool Box > **Spatial Analyst Tools** > **Reclass** > **Reclassify.** Double-Click on Reclassify tool.

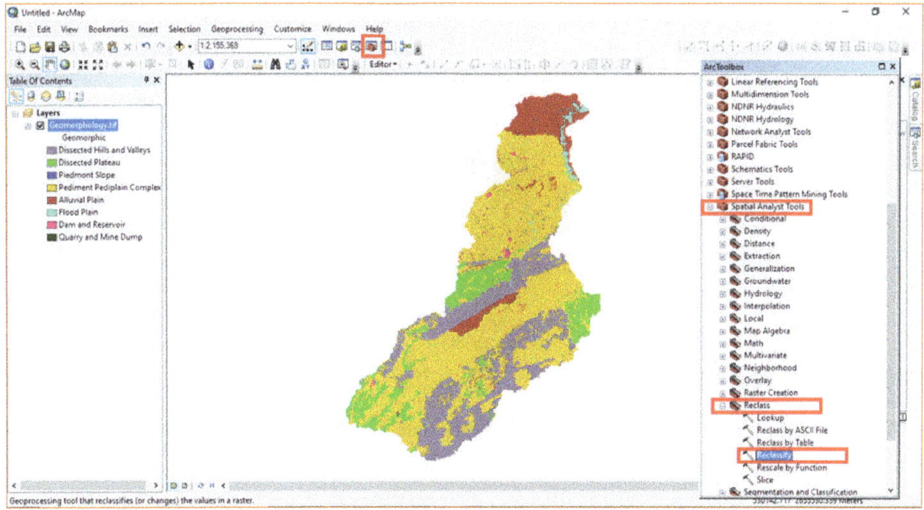

Add 'Geomorphology.tif' file in **Input raster** option, select Geomorphic in **Reclass Field** option, assign ranks to each class in **New Values** field according to Table 1, and give output 'Geomorphology_rank.tif' in **Output raster** option, and click on **OK** option.

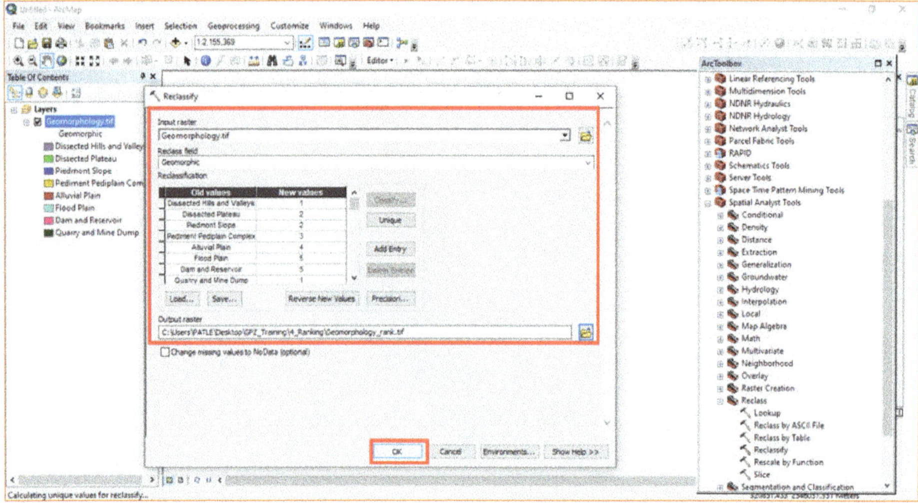

Geomorphology layer has been reclassified according to priority ranks.

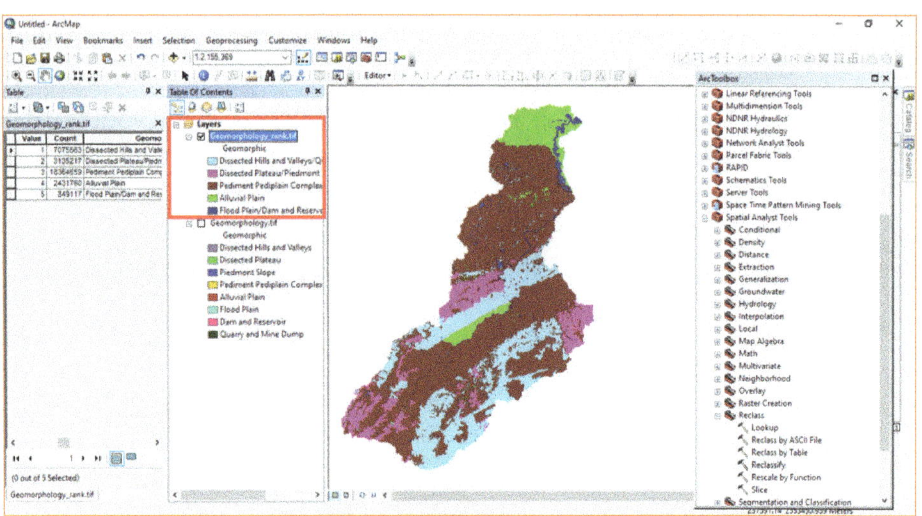

LINEAMENT DENSITY

Open the classified raster layer of Lineament Density in Arc Map application and Click on **ArcTool Box** icon. Then Go to:

ArcTool Box > **Spatial Analyst Tools** > **Reclass** > **Reclassify.** Double-Click on Reclassify tool.

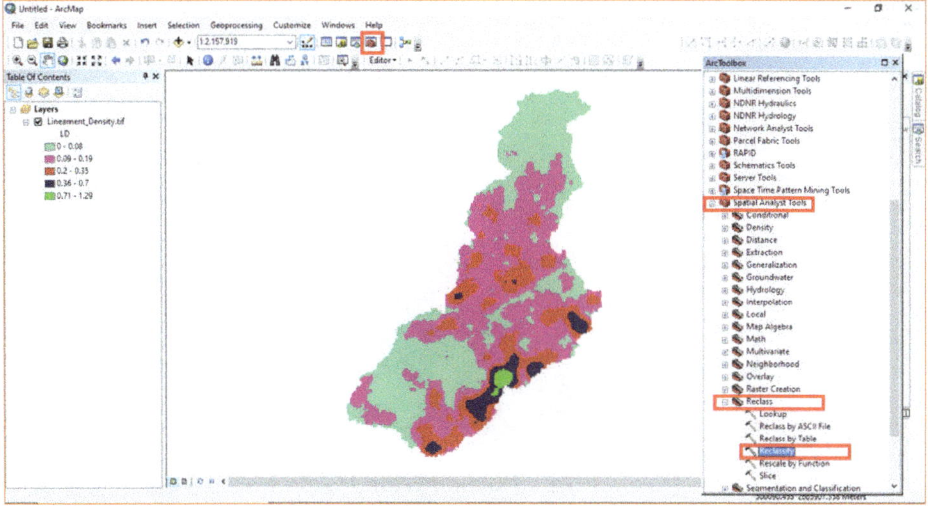

Add 'Lineament_Density.tif' file in **Input raster** option, select LD in **Reclass Field** option, assign ranks to each class in **New Values** field according to Table 1, and give output 'LD_rank.tif' in **Output raster** option, and click on **OK** option.

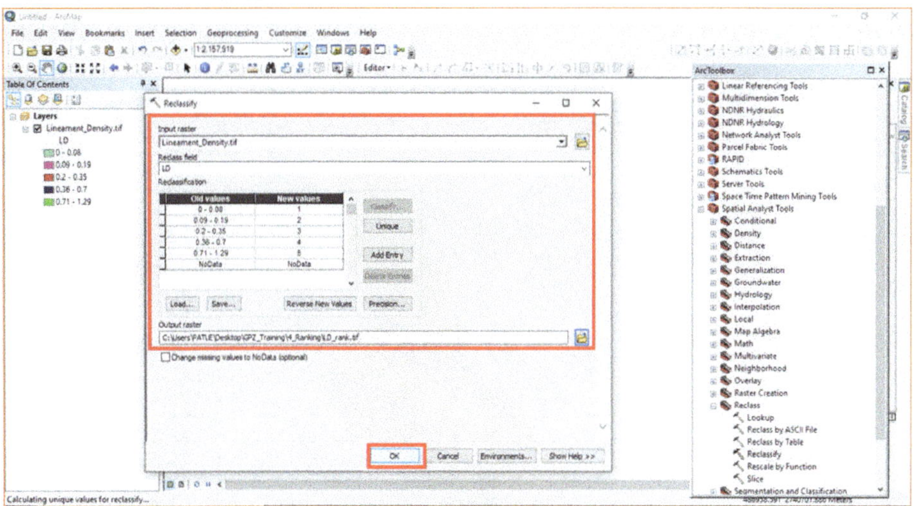

Lineament Density layer has been reclassified according to priority ranks.

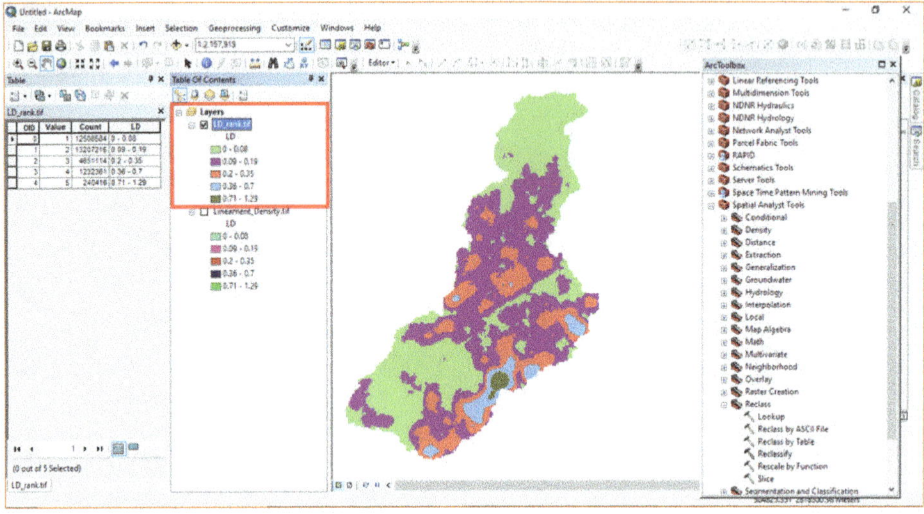

LAND USE/ LAND COVER

Open the classified raster layer of Land Use/ Land Cover in Arc Map application and Click on **ArcTool Box** icon. Then Go to:

ArcTool Box > **Spatial Analyst Tools** > **Reclass** > **Reclassify.** Double-Click on Reclassify tool.

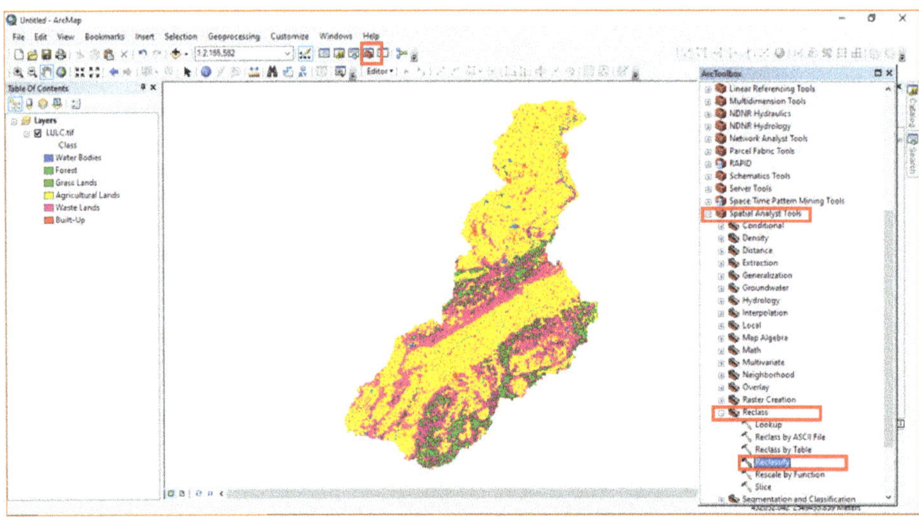

Add 'LULC.tif' file in **Input raster** option, select Class in **Reclass Field** option, assign ranks to each class in **New Values** field according to Table 1, and give output 'LULC_rank.tif' in **Output raster** option, and click on **OK** option.

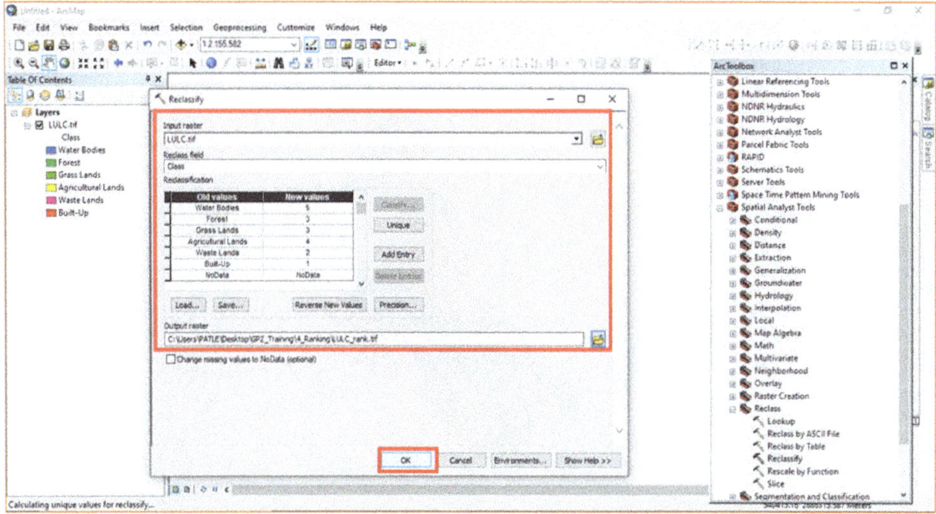

Land Use/ Land Cover layer has been reclassified according to priority ranks.

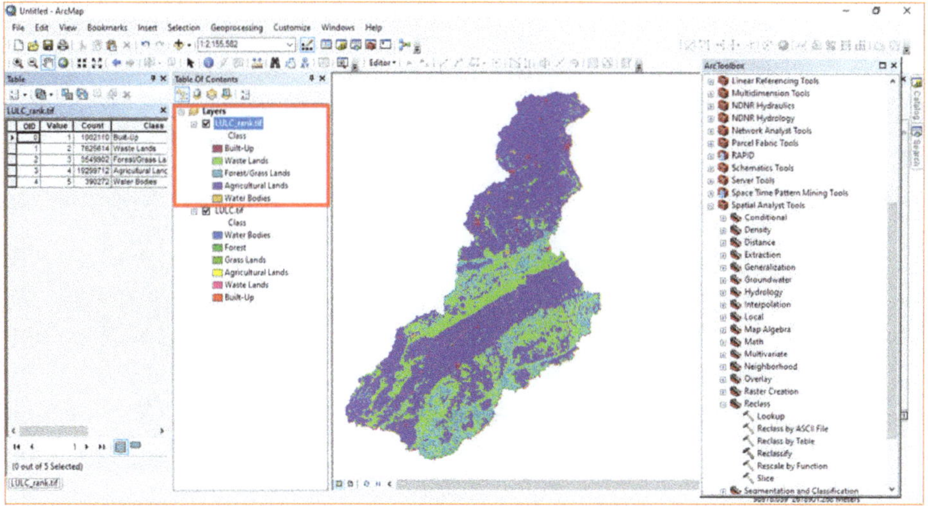

SOIL

Open the classified raster layer of Soil_Texture in Arc Map application and Click on **ArcTool Box** icon. Then Go to:

ArcTool Box > Spatial Analyst Tools > Reclass > Reclassify. Double-Click on Reclassify tool.

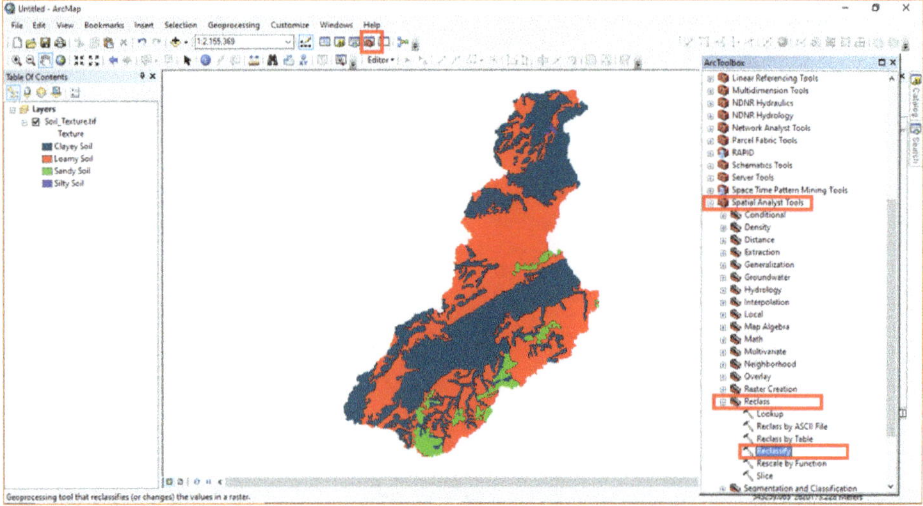

Add 'Soil_Texture.tif' file in **Input raster** option, select Texture in **Reclass Field** option, assign ranks to each class in **New Values** field according to Table 1, and give output 'Soil_rank.tif' in **Output raster** option, and click on **OK** option.

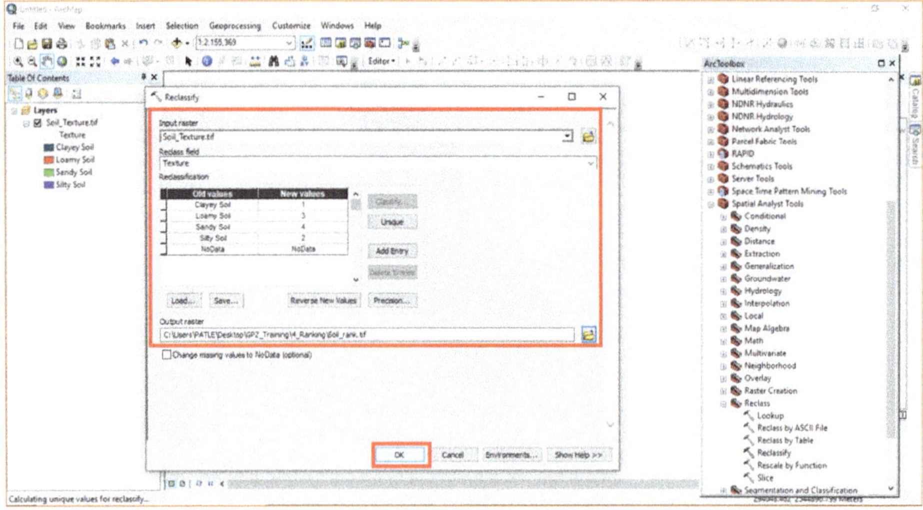

Soil Texture layer has been reclassified according to priority ranks.

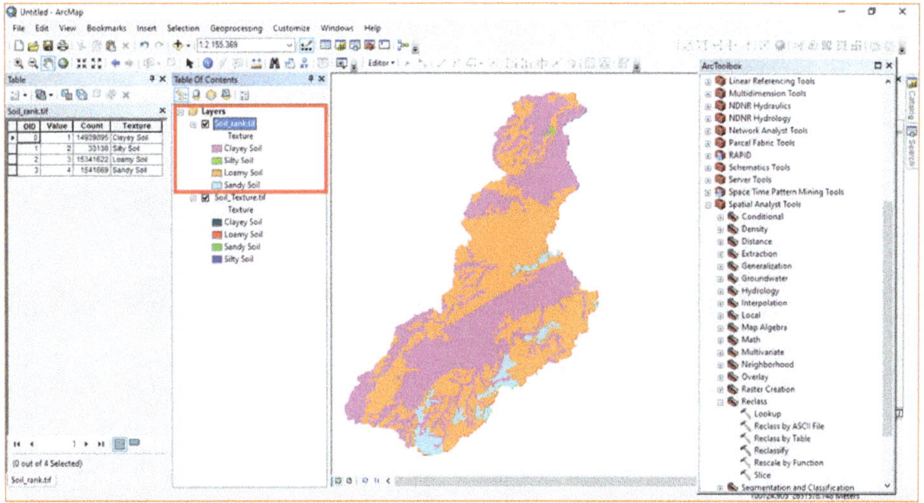

SLOPE

Open the classified raster layer of Slope in Arc Map application and Click on **ArcTool Box** icon. Then Go to:

ArcTool Box > **Spatial Analyst Tools** > **Reclass** > **Reclassify.** Double-Click on Reclassify tool.

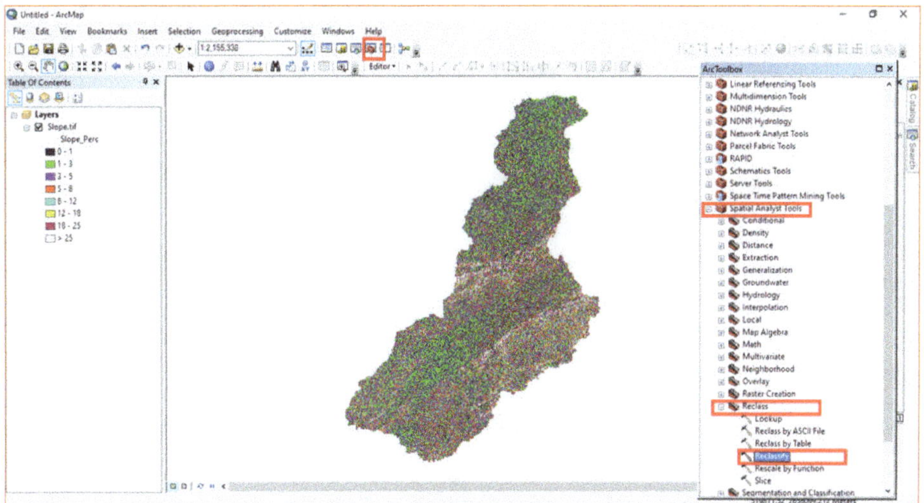

Add 'Slope.tif' file in **Input raster** option, select Slope_Perc in **Reclass Field** option, assign ranks to each class in **New Values** field according to Table 1, and give output 'Slope_rank.tif' in **Output raster** option, and click on **OK** option.

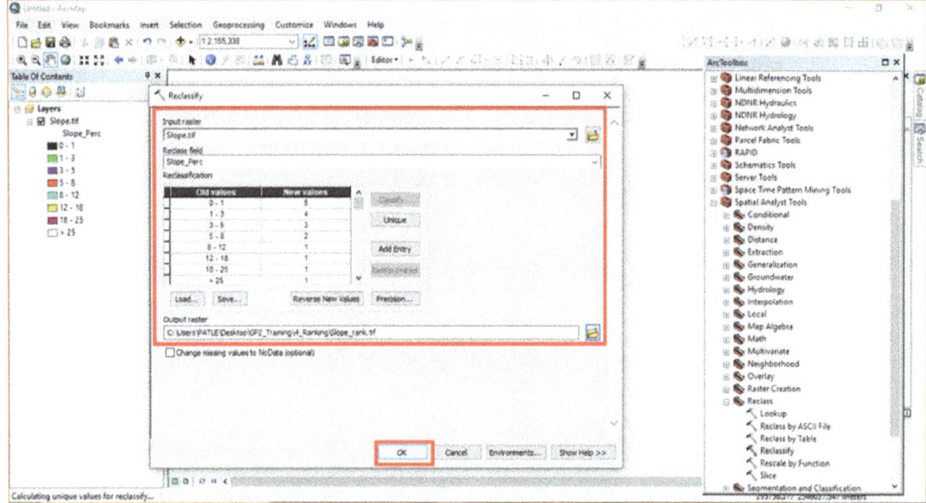

Slope Texture layer has been reclassified according to priority ranks.

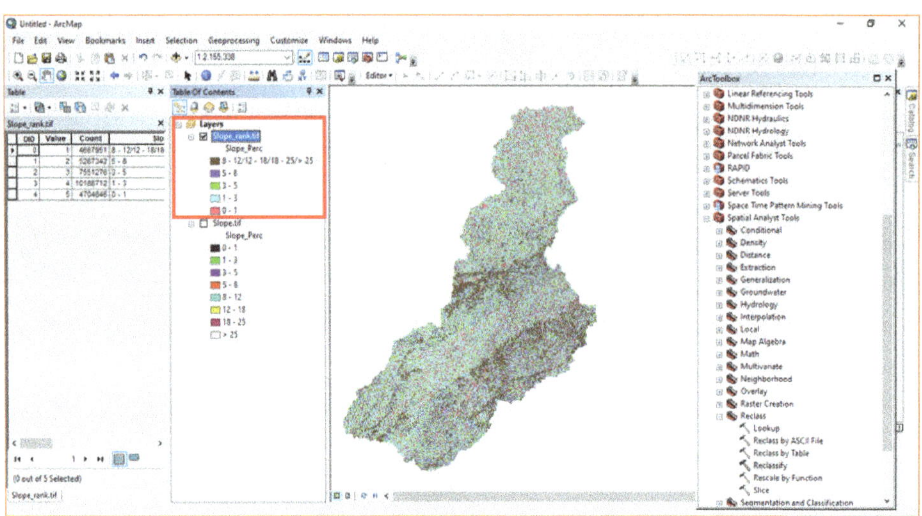

DRAINAGE DENSITY

Open the classified raster layer of Drainage Density in Arc Map application and Click on **ArcTool Box** icon. Then Go to:

ArcTool Box > Spatial Analyst Tools > Reclass > Reclassify. Double-Click on Reclassify tool.

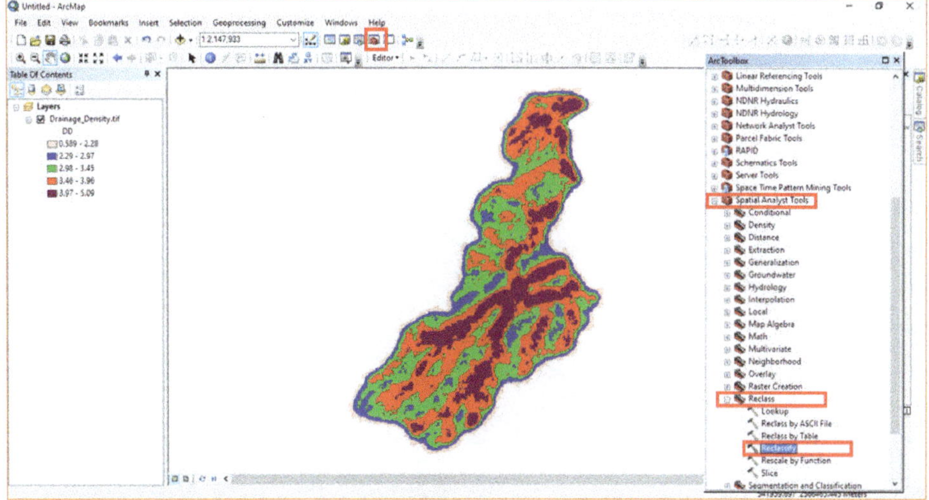

Add 'Drainage_Density.tif' file in **Input raster** option, select DD in **Reclass Field** option, assign ranks to each class in **New Values** field according to Table 1, and give output 'DD_rank.tif' in **Output raster** option, and click on **OK** option.

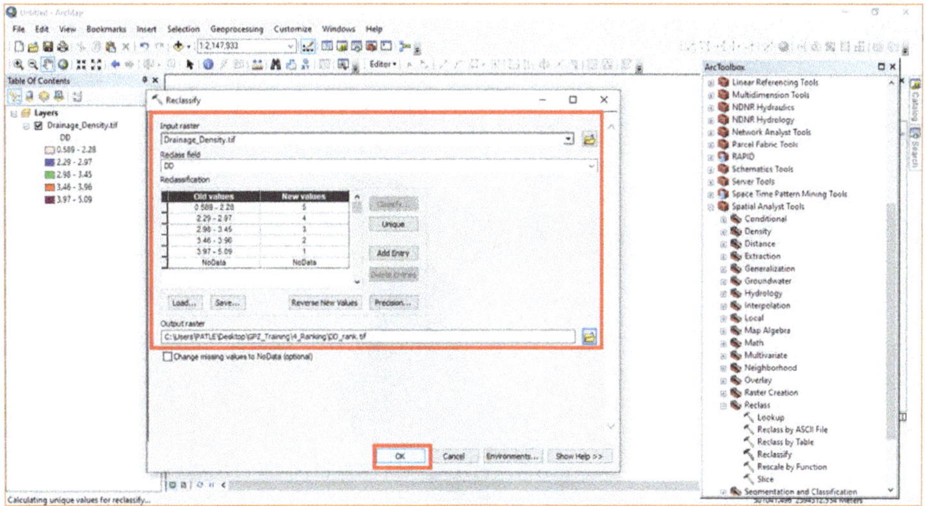

Drainage Density layer has been reclassified according to priority ranks.

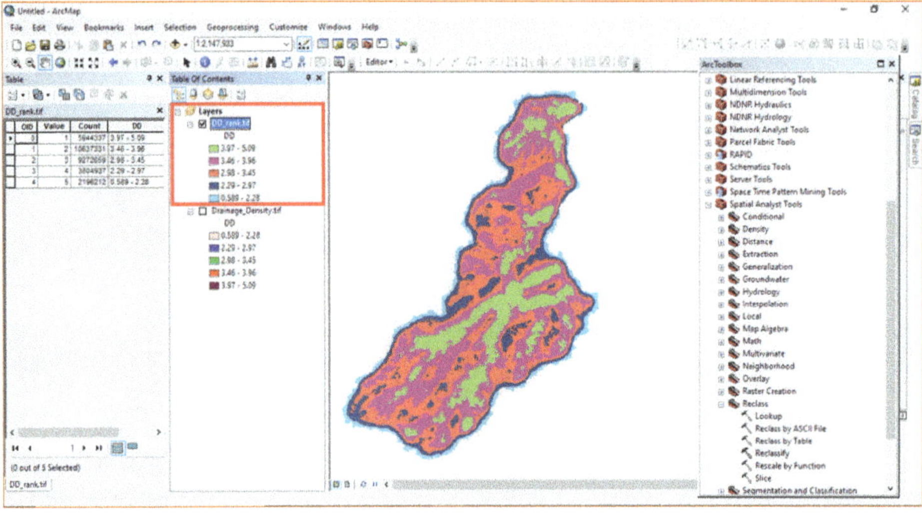

RAINFALL

Open the classified raster layer of Rainfall in Arc Map application and Click on **ArcTool Box** icon. Then Go to:

ArcTool Box > **Spatial Analyst Tools** > **Reclass** > **Reclassify.** Double-Click on Reclassify tool.

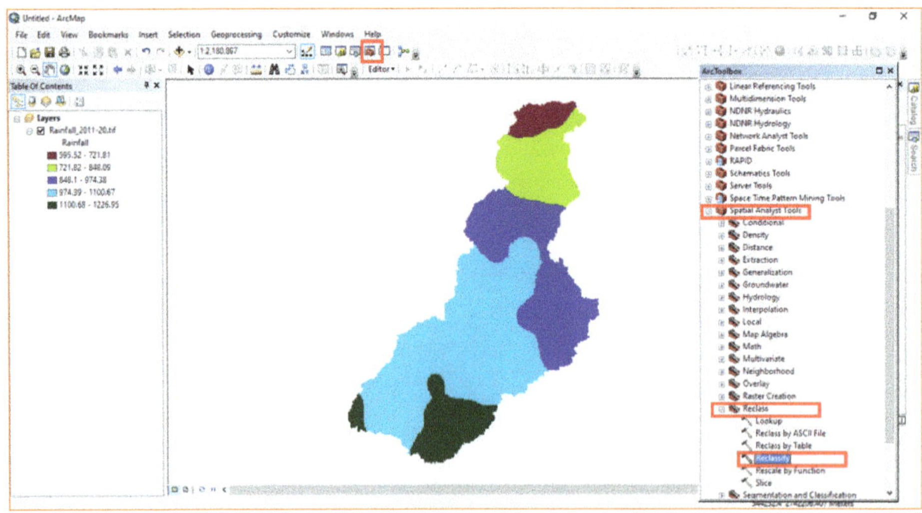

Add 'Rainfall_2011-20.tif' file in **Input raster** option, select Rainfall in **Reclass Field** option, assign ranks to each class in **New Values** field according to Table 1, and give output 'Rainfall_rank.tif' in **Output raster** option, and click on **OK** option.

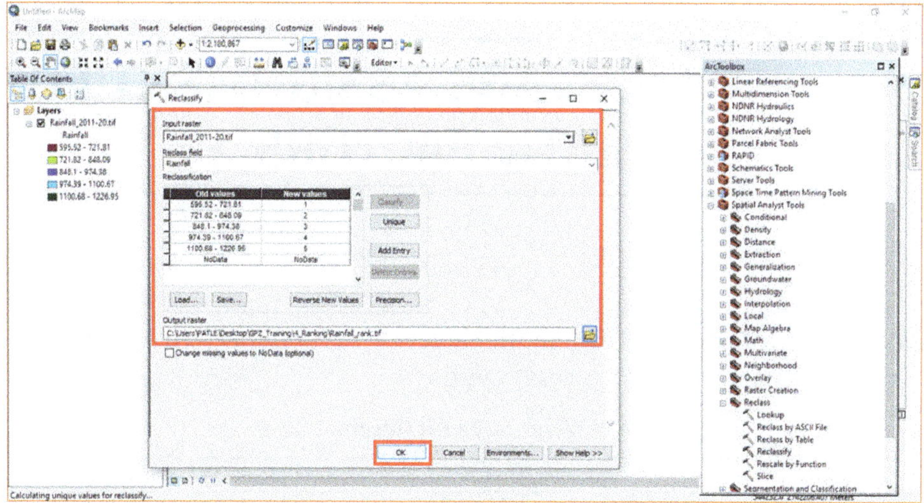

Drainage Density layer has been reclassified according to priority ranks.

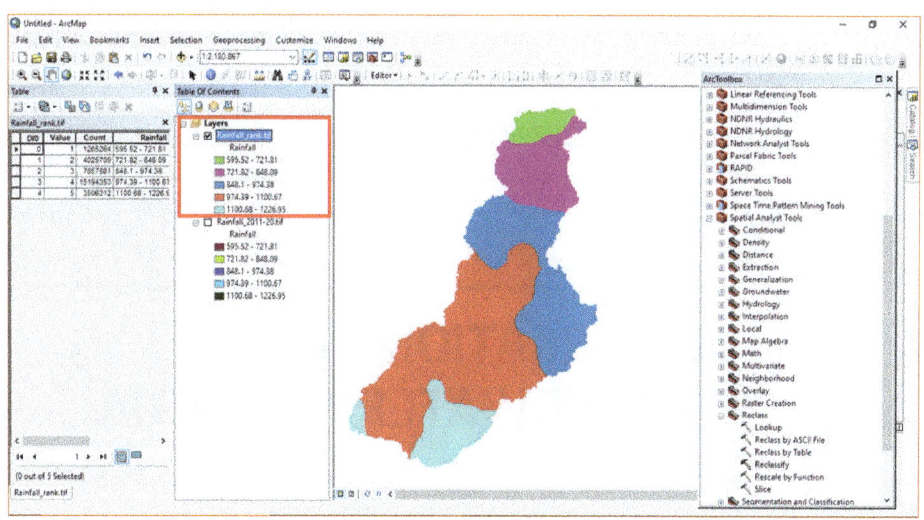

WEIGHTAGE ASSIGNMENT FOR THEMATIC FACTORS

Weightages of different thematic layers are obtained from the AHP (Analytical Hierarchical Process) method.

Table 2: Weightages for different layers

S. No.	Factors	Weightage (%)
1.	Geology	22
2.	Geomorphology	20
3.	Lineament Density (km/ km^2)	15
4.	Land Use/ Land Cover	12
5.	Soil Texture	7
6.	Slope (%)	9
7.	Drainage Density (km/ km^2)	8
8.	Rainfall (mm)	6
Total		100

DEMARCATION OF GROUNDWATER POTENTIAL ZONES

All these parameters are integrated in ArcGIS 10.8 software and the resultant map is created through the raster calculation by using equation employed in weighted sum.

$$GPZ = \begin{bmatrix} GL_W \cdot GL_R + GM_W \cdot GM_R + LD_W \cdot LD_R + LULC_W \cdot LULC_R \\ + ST_W \cdot ST_R + + RL_W \cdot RL_R + SL_W \cdot SL_R + DD_W \cdot DD_R \end{bmatrix}$$

... (Equation 1)

Where, GPZ is the groundwater potential zones, W is weight of each factor, R is rank of each class, GL is the geology, GM is geomorphology, LD is lineament density, LULC is the land use land cover, ST is soil texture, RL is the rainfall, SL is slope, and DD is drainage density

DEMARCATION OF GROUNDWATER POTENTIAL ZONES

Browse the 'Ranking' folder, select on the ranked raster layers, and then click on **Add** option.

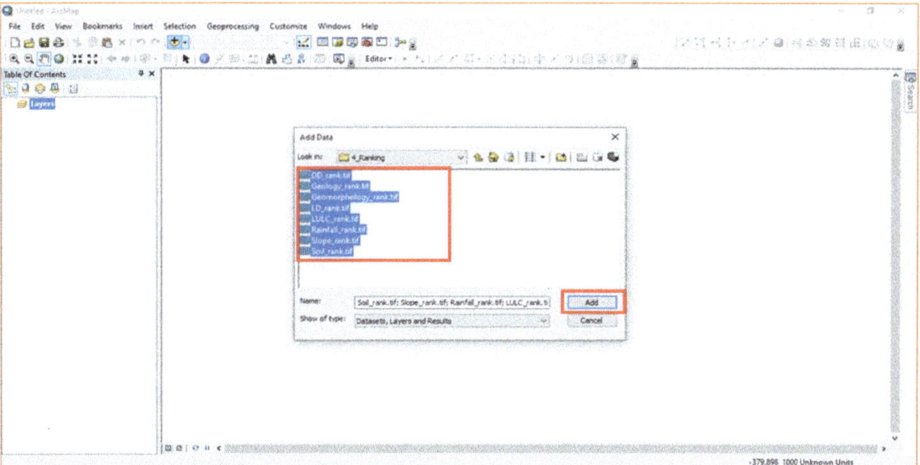

Click on **ArcTool Box** icon. Go to:

ArcTool Box > Spatial Analyst Tools > Map Algebra > Raster Calculator. Double-Click on **Raster Calculator** tool.

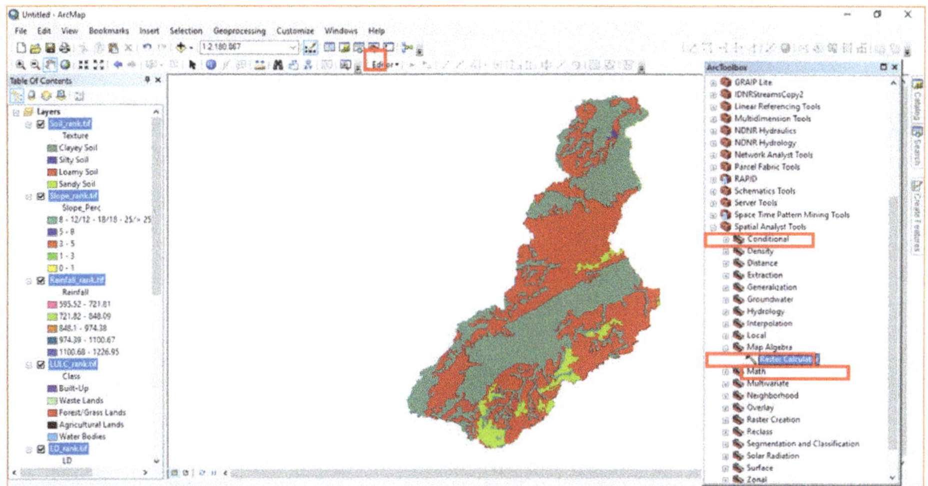

Dialog Box of **Raster Calculator** has been opened.

Click on layer and multiply with its weightage. Apply same for all the layers and summed up.

The Expression can be written as:

("Geology_rank.tif" * 22) + ("Geomorphology_rank.tif" * 20) + ("LD_rank.tif" * 15) + ("LULC_rank.tif" * 12) + ("Soil_rank.tif" * 7) + ("Slope_rank.tif" * 9) + ("DD_rank.tif" * 8) + ("Rainfall_rank.tif" * 6)

Give the Output name 'GPZ.tif' in **Output Raster** option and click on **OK** button.

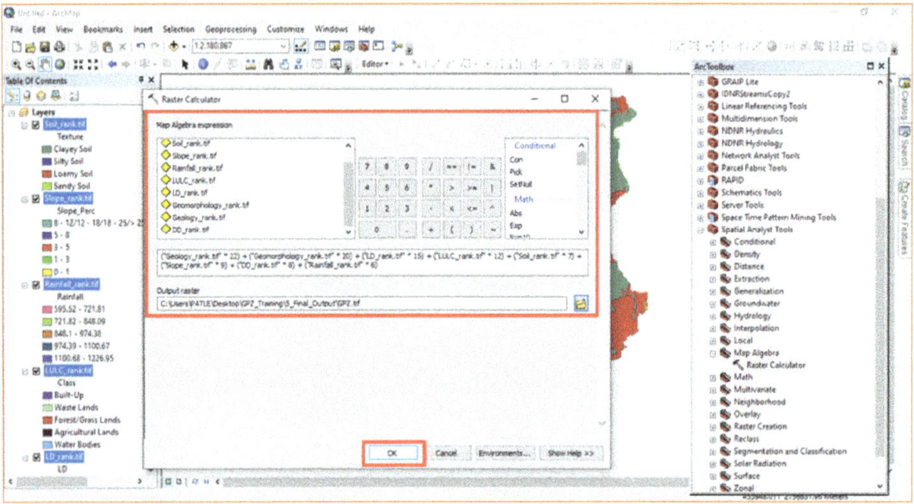

Final output of Groundwater Potential Map has been Generated.

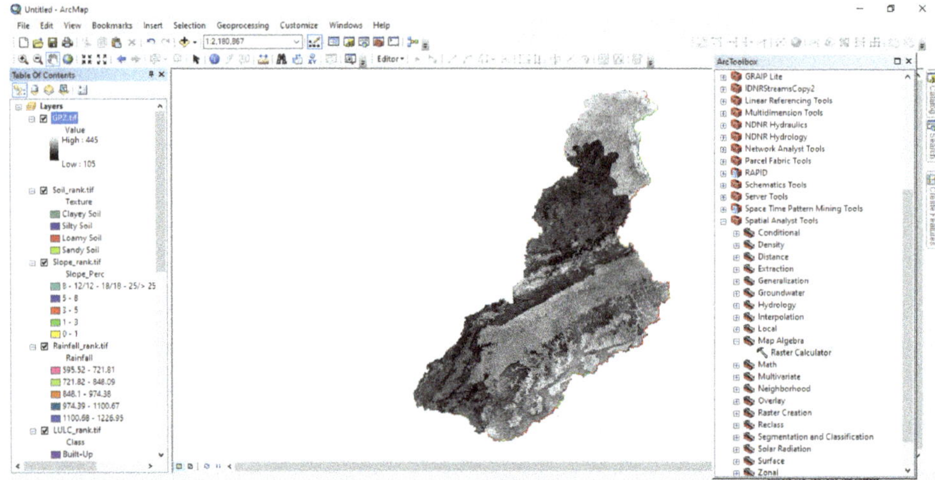

Generated GPZ.tif file is available in the form of dynamic raster layer.

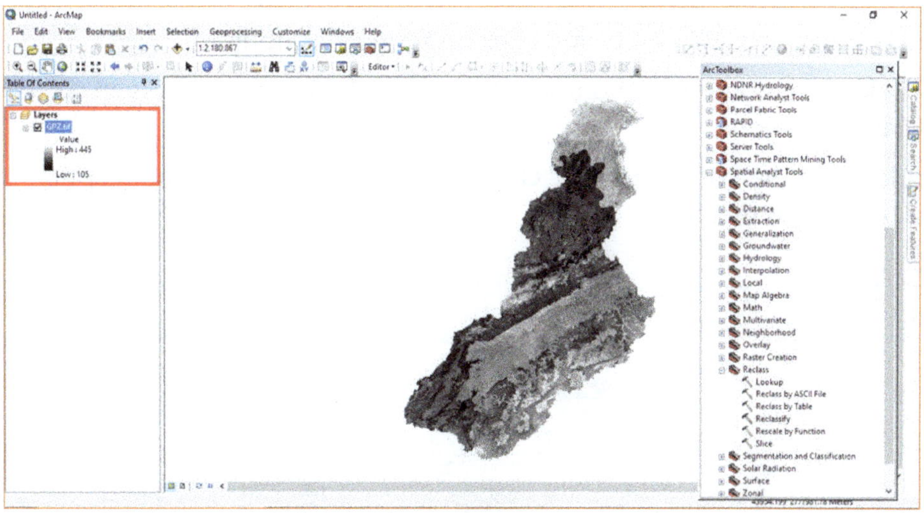

The obtained integrated layer ranged from 105 to 445 index value which was classified into five classes viz. < 230, 230 - 260, 260 - 290, 290 - 320 and > 320 index value of groundwater potential map.

Less index value reflects low groundwater potential and higher index value represents the high groundwater potential of an area.

Groundwater potential zones are classified into five different classes namely very good, good, moderate, poor, and very poor with respect to high to low index values.

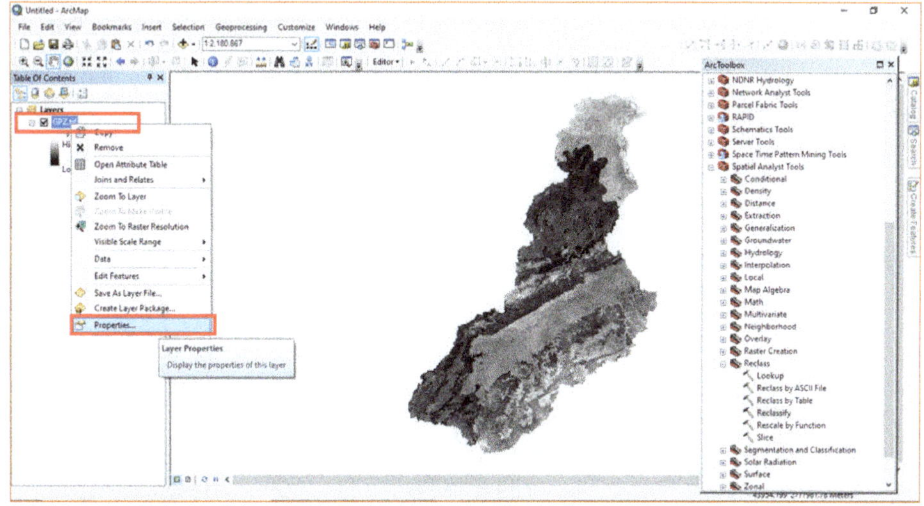

The raster map of Groundwater Potential Zones needs to reclassify according to index values.

Right click on **GPZ.tif** file and click on **Properties** option.

Click on **Symbology** option, select the **Classified** option. and then click on **Classify** option.

Select **number of classes** in class option and select **Manual** option in classification method.

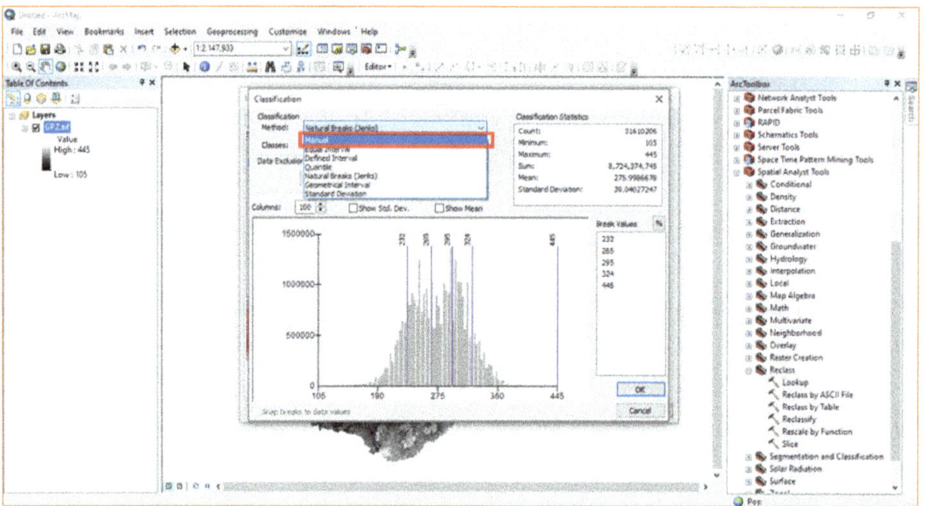

Go to **Breaks Values** % panel, give the break values <230 to 445 with equal interval 30, and click on **OK** option.

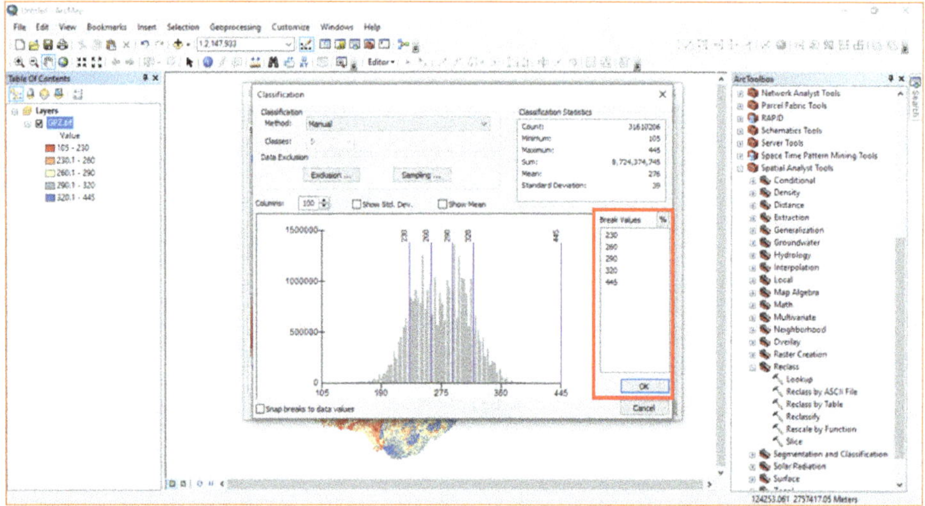

Change the color from **Color Ramp**, click on **Apply** option, and hit the **OK** option.

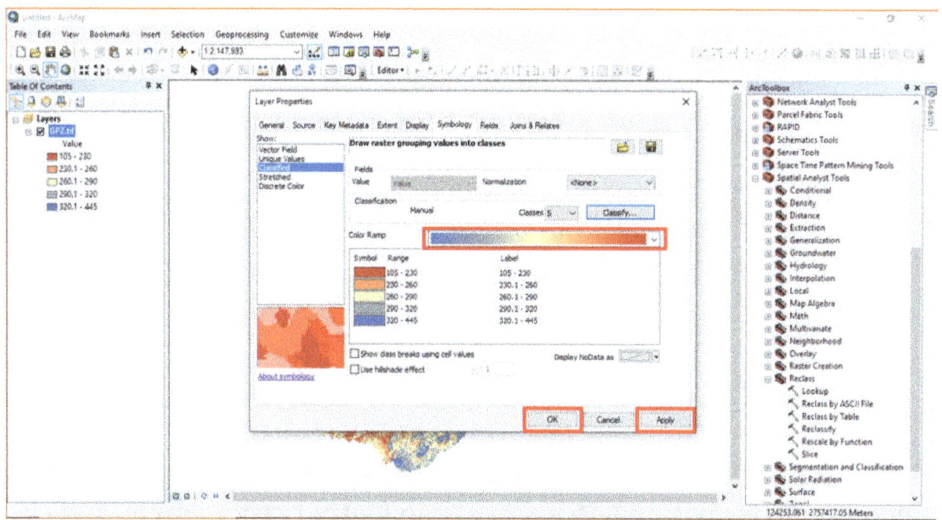

GPZ map of Ken Basin has been classified in five Classes.

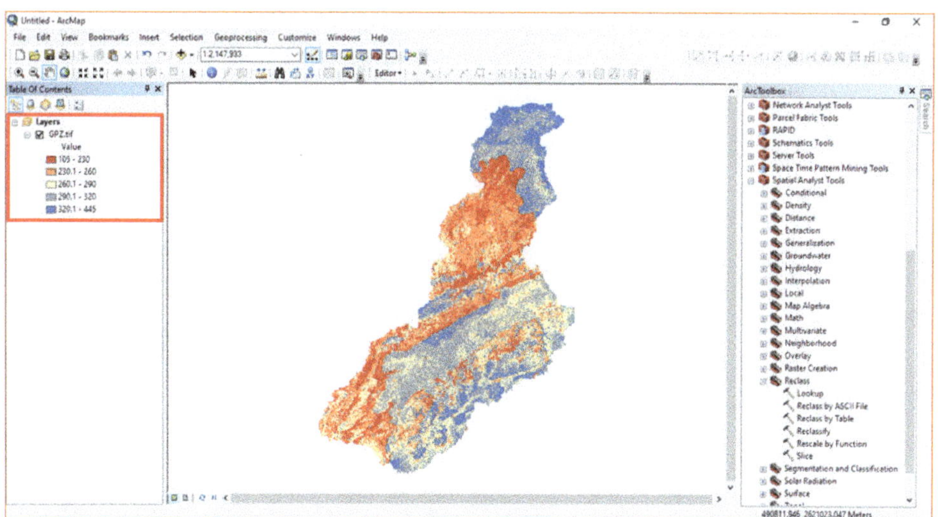

Go to:

ArcTool Box > Spatial Analyst Tools > Reclass > Reclassify. Double-Click on Reclassify tool.

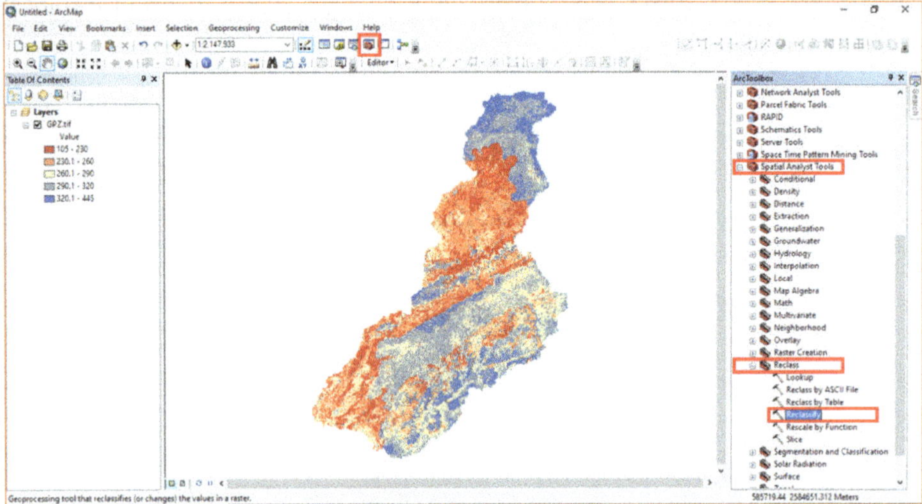

Add 'GPZ.tif' file in **Input raster** option, select value in **Reclass Field** option, and give output GPZs_Class.tif in **Output raster** option, and click on **OK** option.

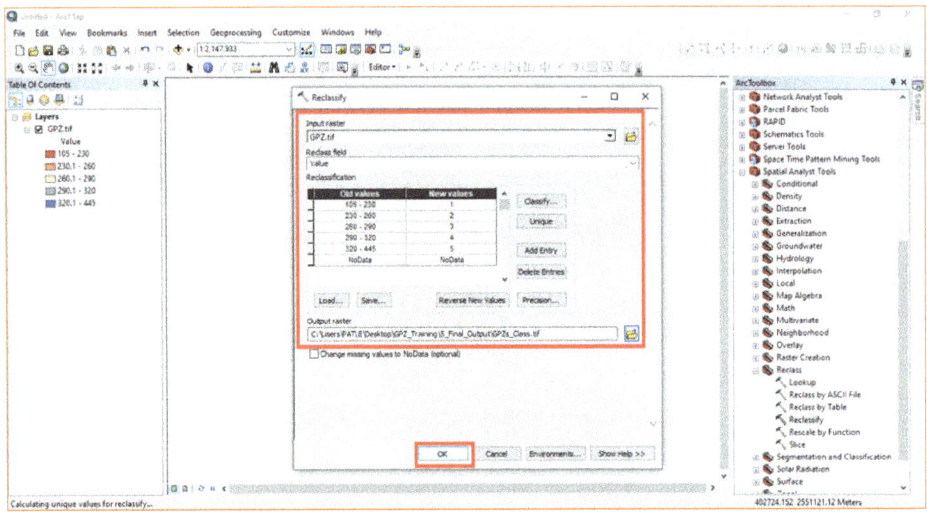

GPZ map has been reclassified.

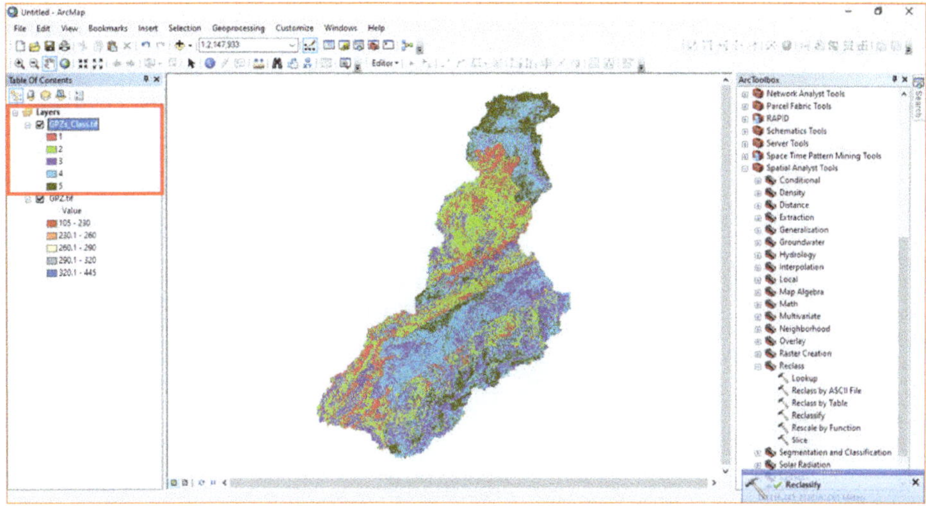

Right click on reclassified layer and click on **Open Attribute Table** option, and click on **Table Option**.

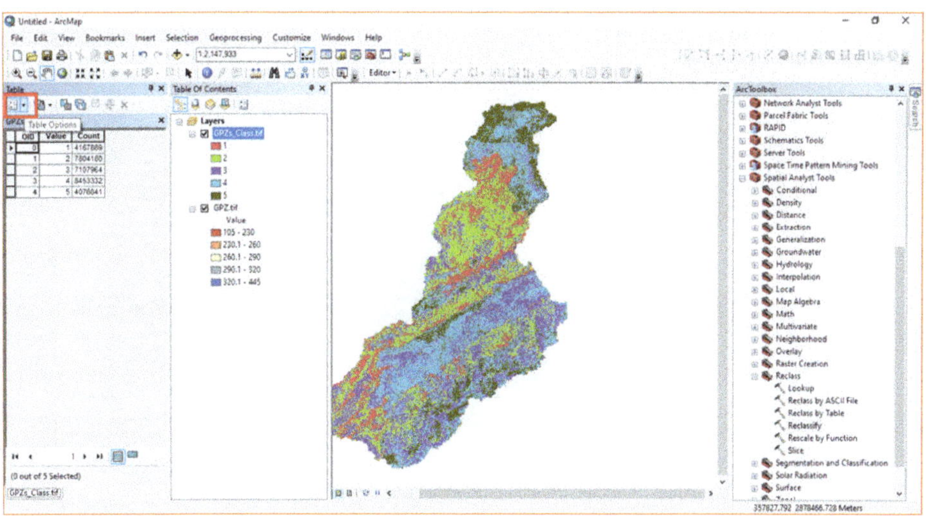

Click on **Add Field** option, give **field name** is GPZ_Class and type select as **text**, and hit the **OK** button.

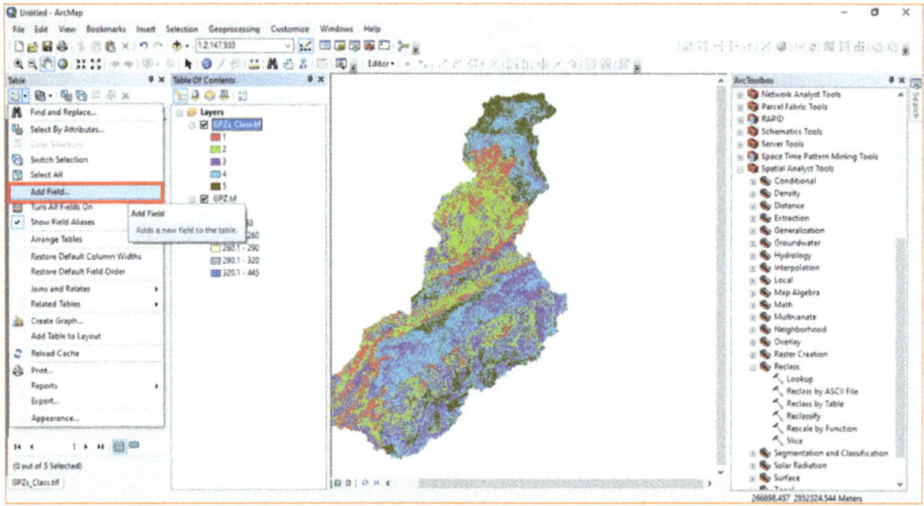

Again, right click on reclassified layer and click on **Open Attribute Table** option. Then, go to **Editor** menu and click on **Start Editing** option.

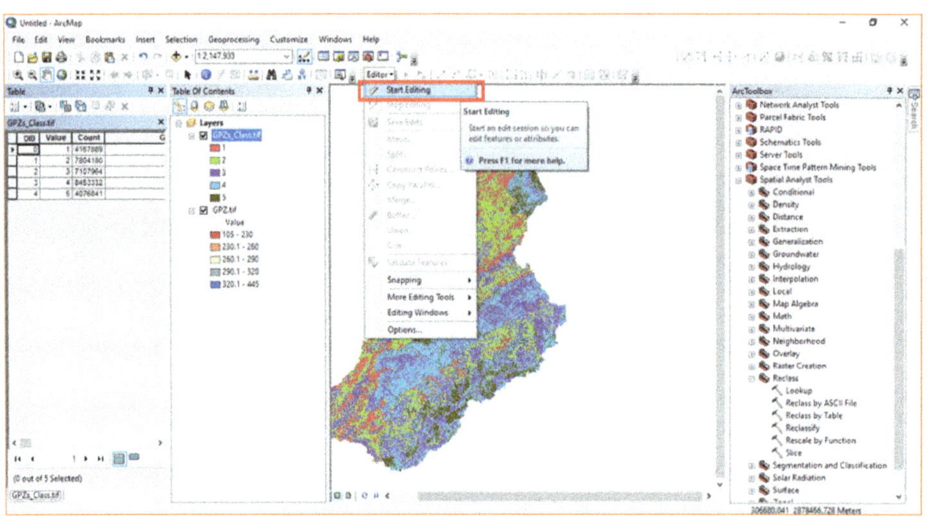

Go to **GPZ_Class** field, fill the Class name of Groundwater Potential Zones, and click on **Save Edits** option.

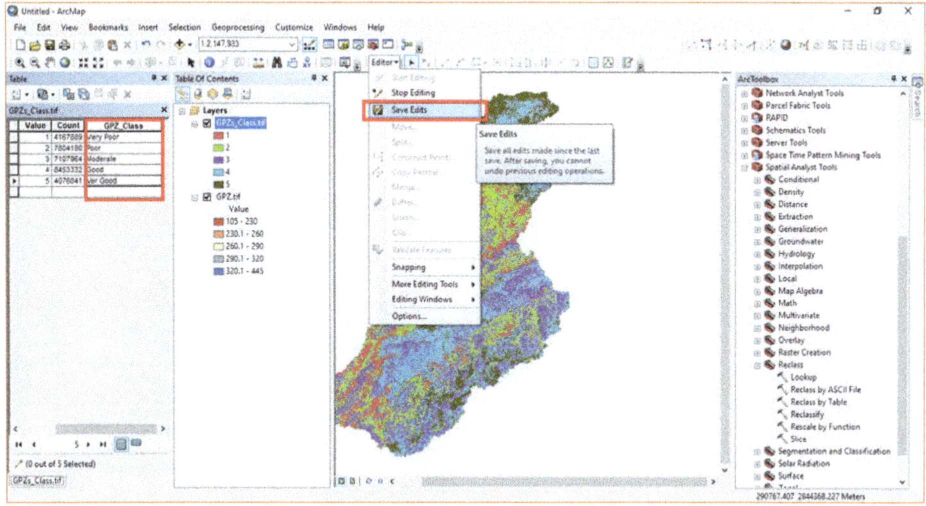

Click on **Stop Editing** option.

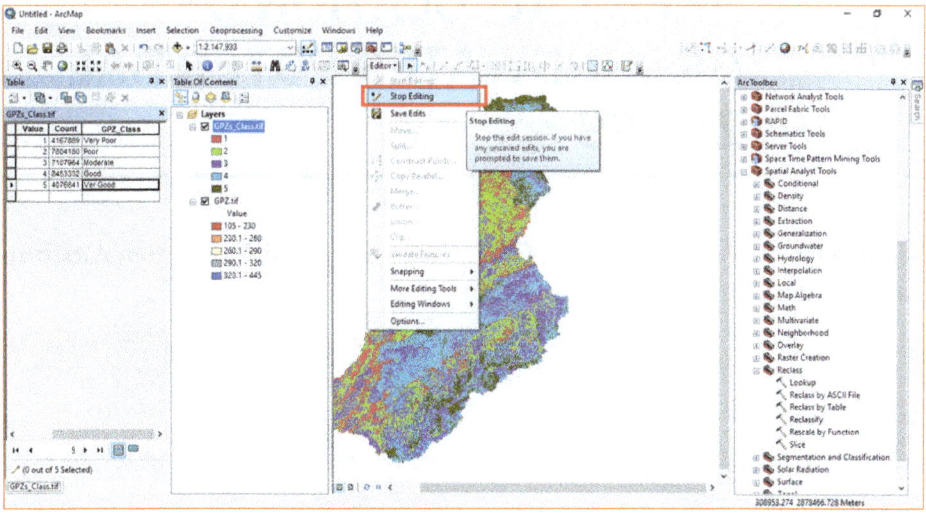

Groundwater Potential Zones Map of Ken Basin has been Properly classified. The color ramp has been changed to better understanding from Very Good to Very Poor classes of GPZ.

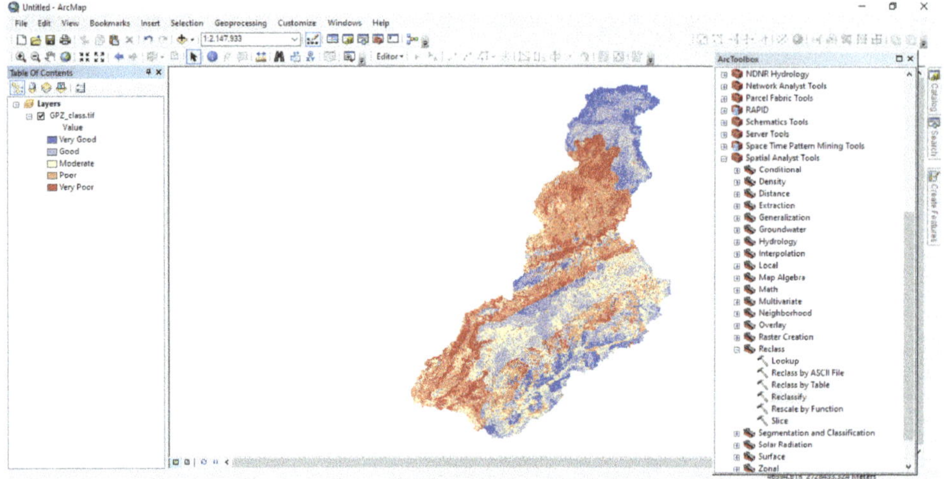

AREA CALCULATION OF GROUNDWATER POTENTIAL ZONES

Right click on reclassified layer **GPZ_Class.tif** and click on **Open Attribute Table** option, and click on **Table Option**.

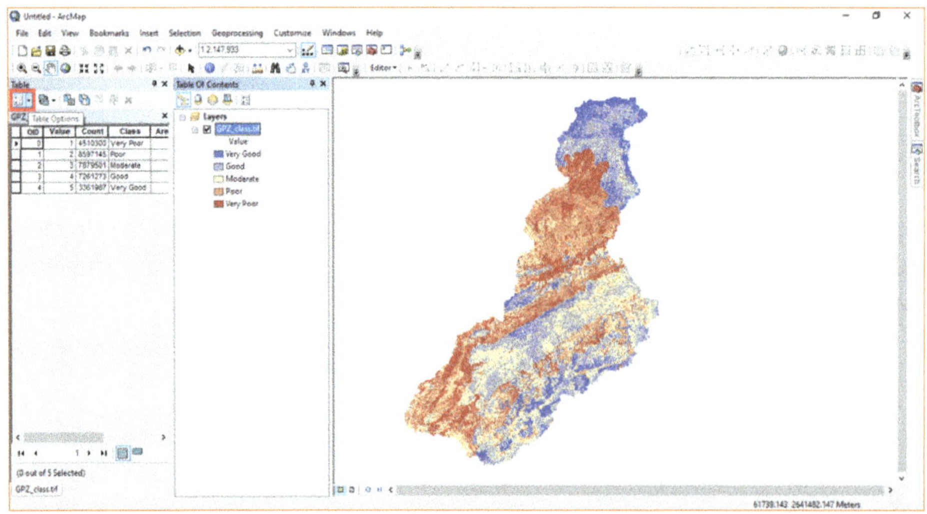

Click on **Add Field** option.

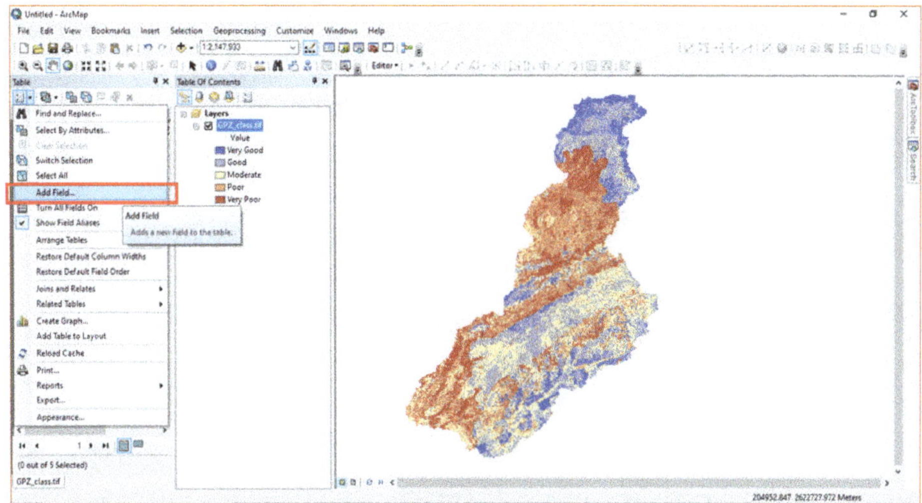

Give **field name** is Area_km2 and type select as **Float**, and hit the **OK** button.

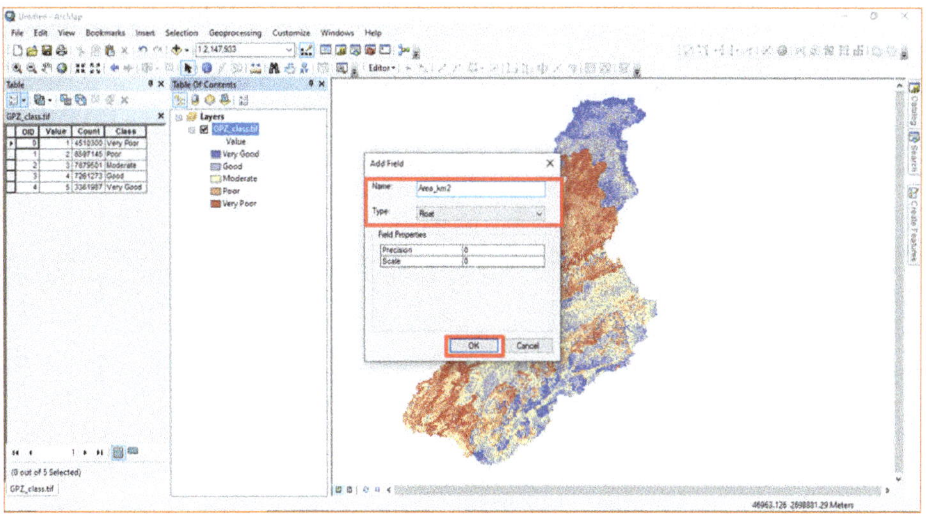

Right Click on **Area_km2** field and then click on **Field Calculator** option.

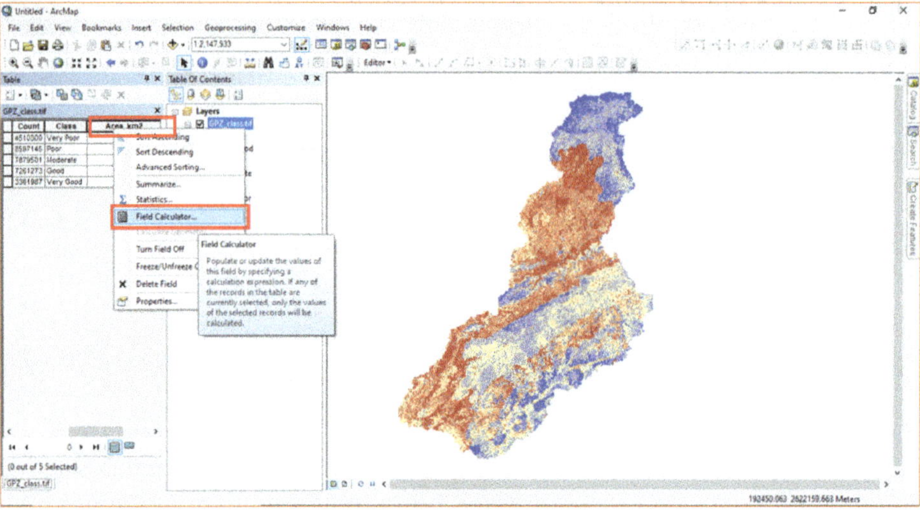

Double-Click on **Count** option from **Fields** panel and type the expression given as below:

[Count] *30*30/1000000]

In above expression, value 30 represents the cell size or spatial resolution of the raster data.

Then, click on the **OK** option.

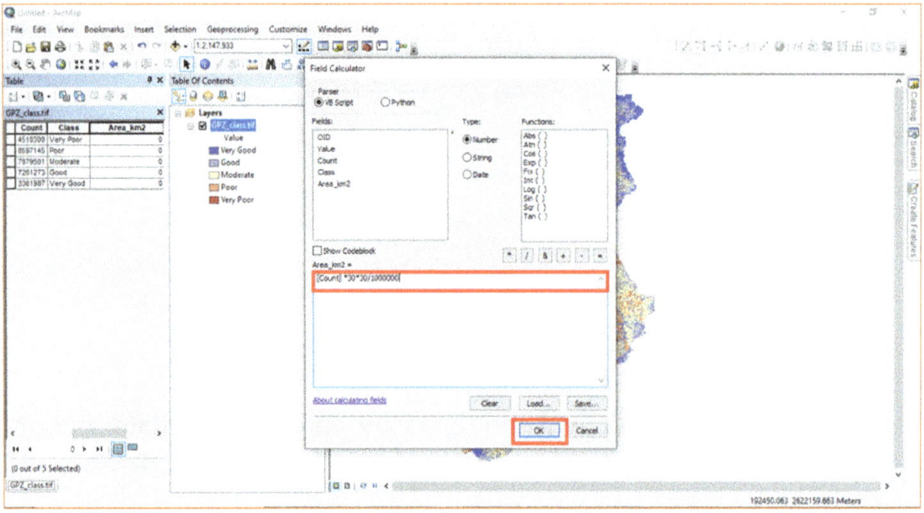

Area of different GPZ classes has been calculated in square kilometers (sq. km).

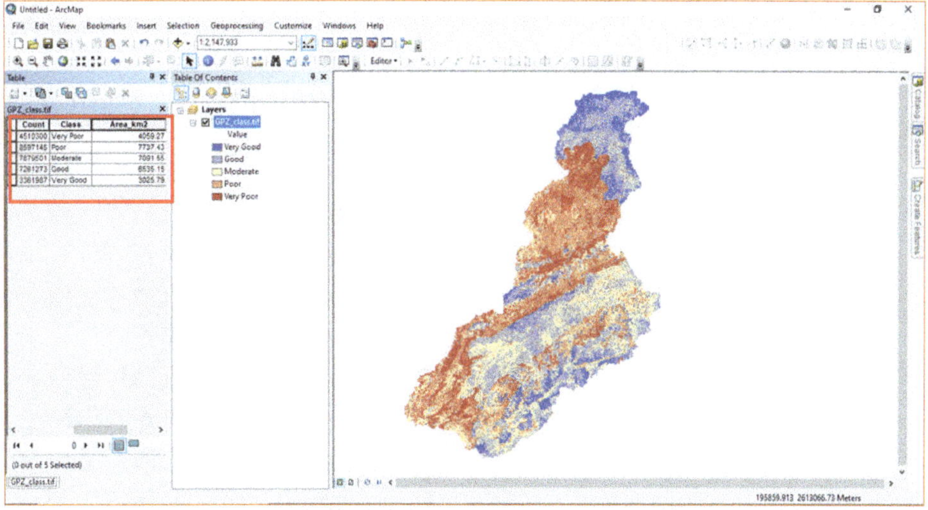

VALIDATION OF GROUNDWATER POTENTIAL ZONES

Firstly, add the Groundwater Potential Zones map in **Arc Map** software.

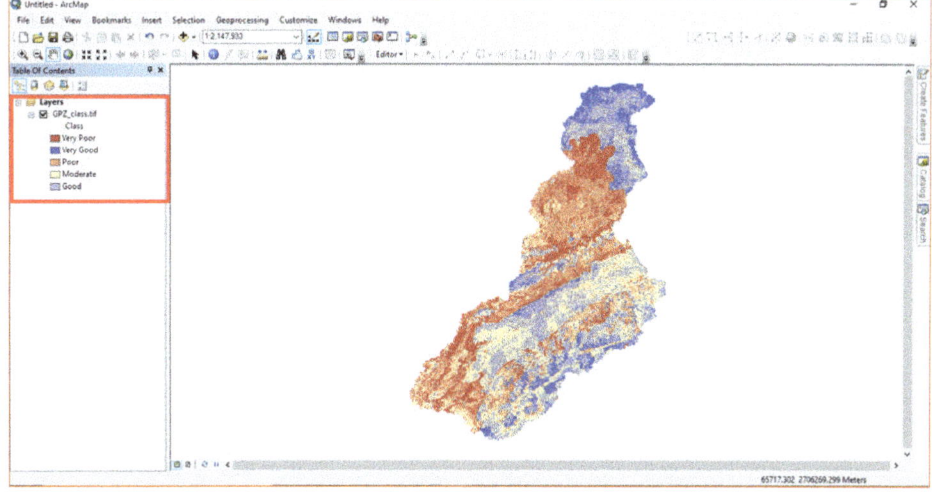

Click on **Arc Toolbox** option.

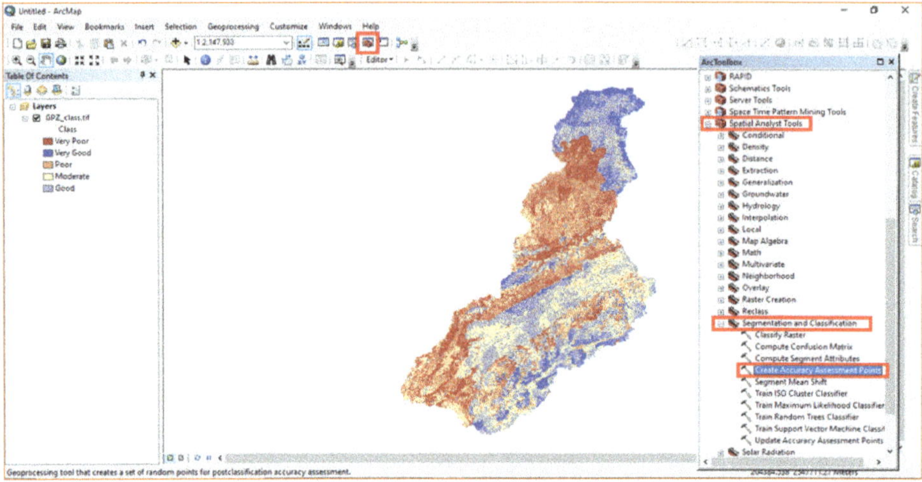

Arc Toolbox > Spatial Analyst Tools > Segmentation and Classification > Create Accuracy Assessments Points

Double-Click on **Create Accuracy Assessments Points** tool.

Select 'GPZ_Class.tif' file in **Input raster or feature class data** option, give the output name 'Random_Points_100.shp' in **Output Accuracy Assessments points** option, and choose the 'STRATIFIED_RANDOM' method in **Sampling Strategy** option, and hit the **OK** button.

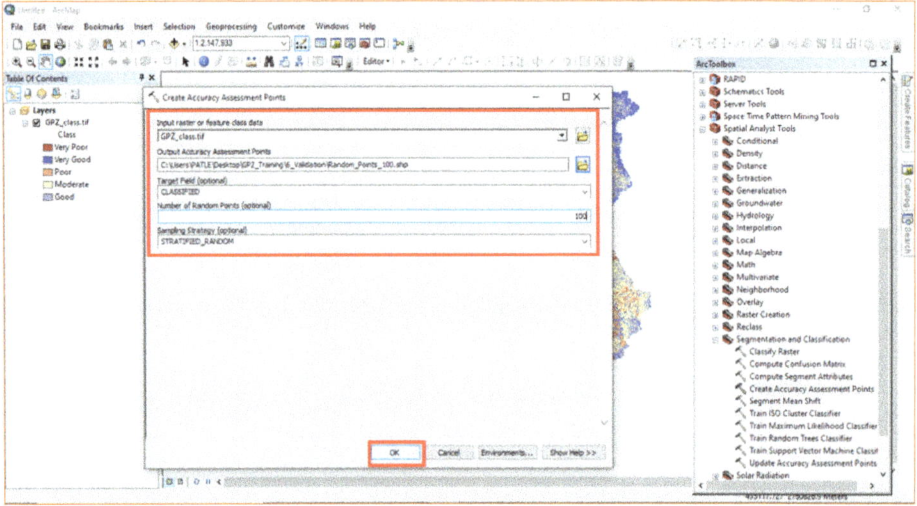

Random Points (accuracy assessment points) has been generated.

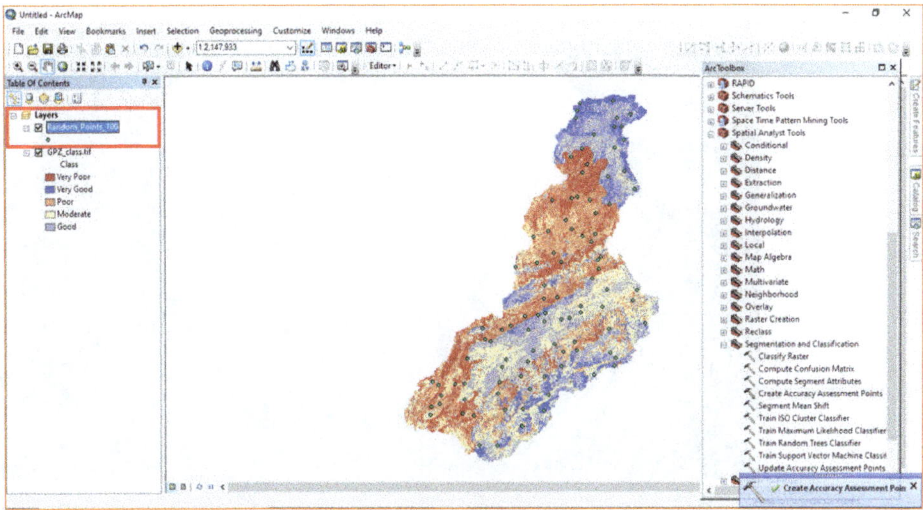

Right-Click on file name 'Random_Points_100' and click on the **Open Attribute Table** option.

Classified column represents the Classes of GPZ (1 to 5: Very Poor to Very Good) and **GrndTruth** column needs to be modified after the verification of each class.

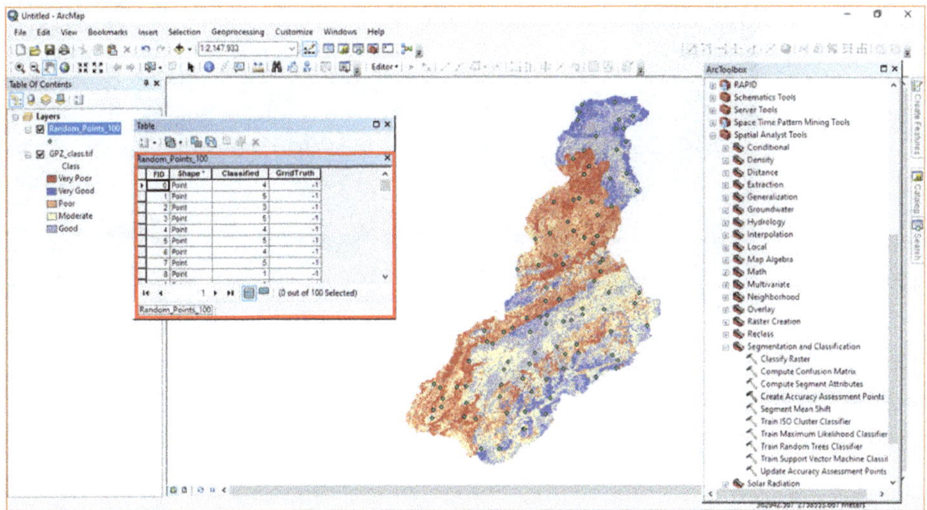

Open the **Google Earth Pro** software.

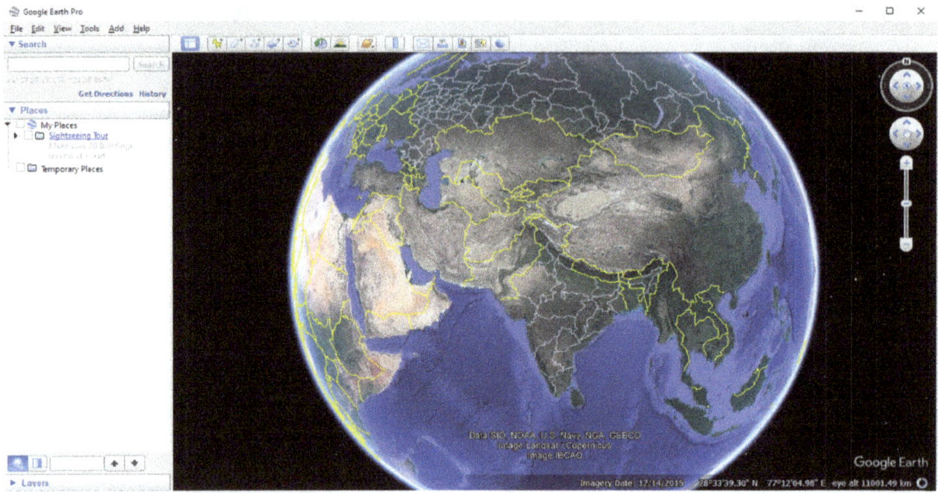

Click on **Import** option.

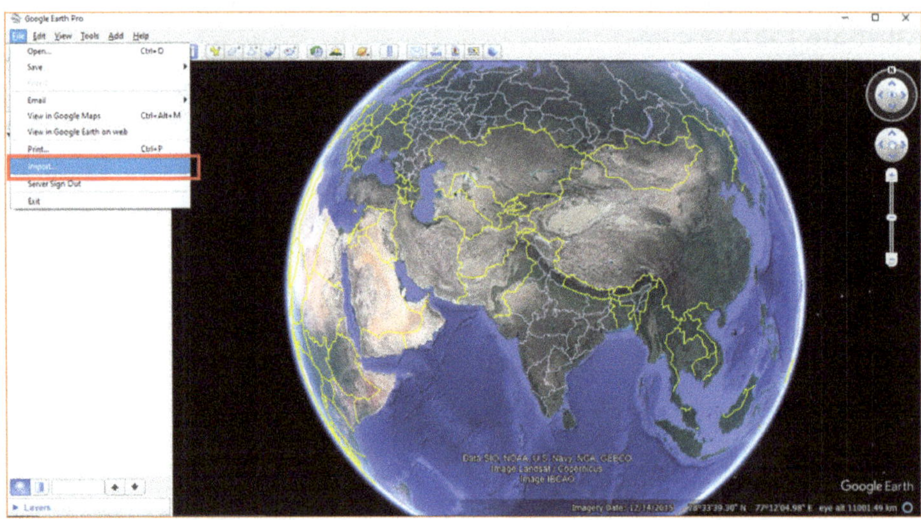

Browse the folder having shapefile of Random Points and Select ESRI Shape (*.shp) format.

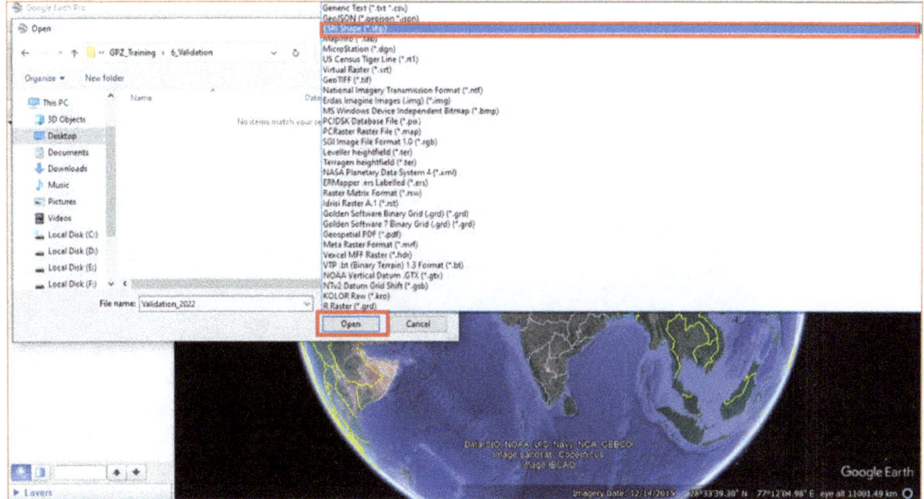

Select the 'Random_Points_100' file and click on **Open** option.

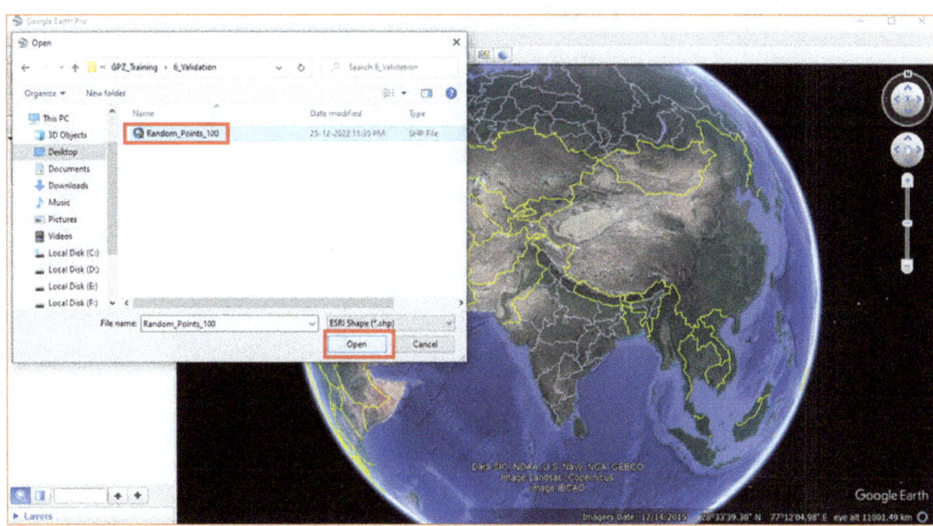

Random points file has been imported in Google Earth Pro.

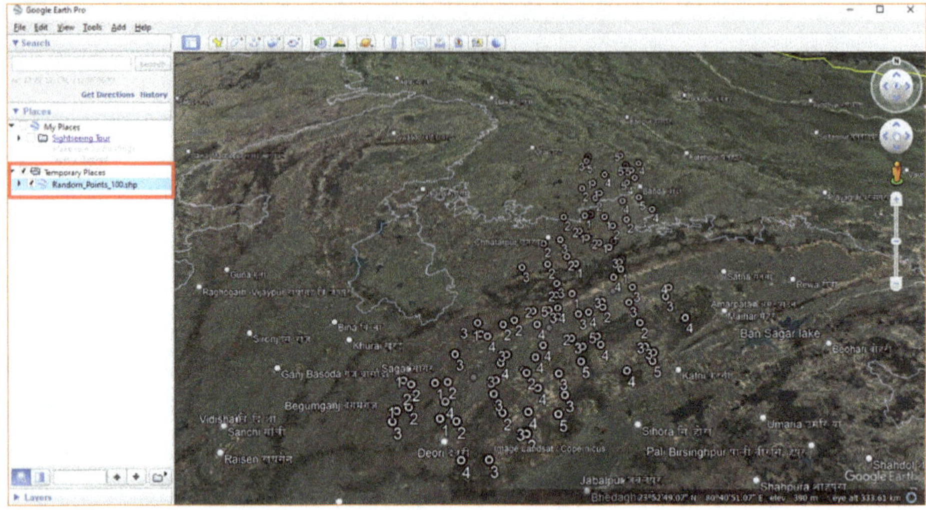

Double-Click on 'Random_Points_100.shp' file. All the points containing GPZ class is showing in **Places** panel.

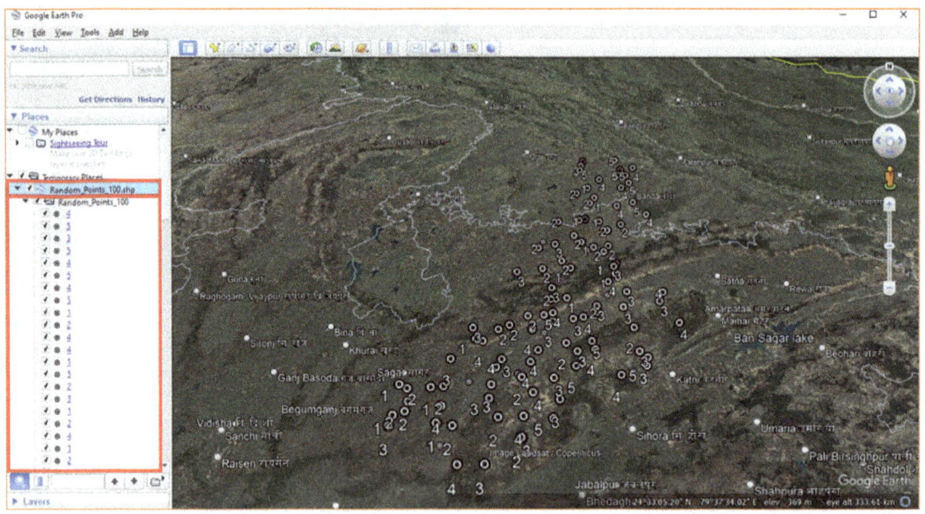

Right-Click on File name, click on **Save Places As** option, and then browse the target folder.

Give file name **Random_Points_100**, choose Kml (*.kml) format in **Save as type** option, and click on **Save** option.

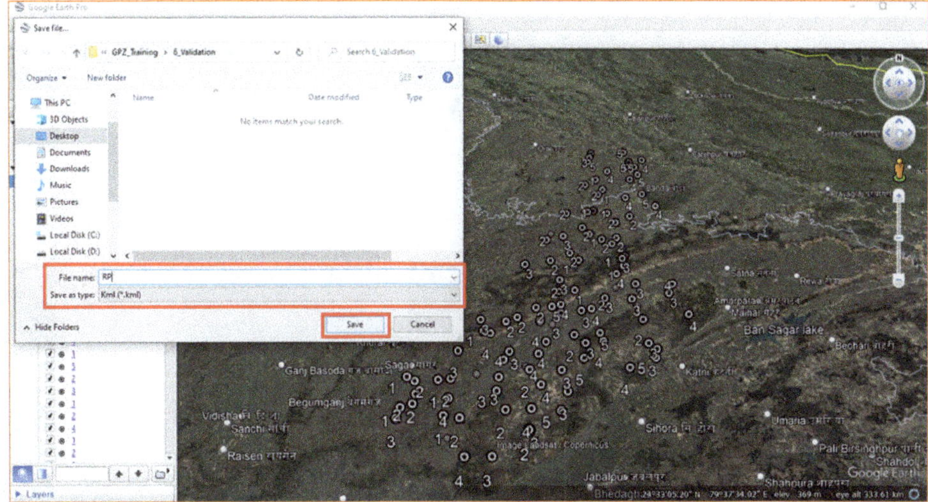

Click on First point, Only Classified and GrndTruth option is showing. So, there is a need to fill the information manually of particular point.

Right click on it, and click on **Properties** option.

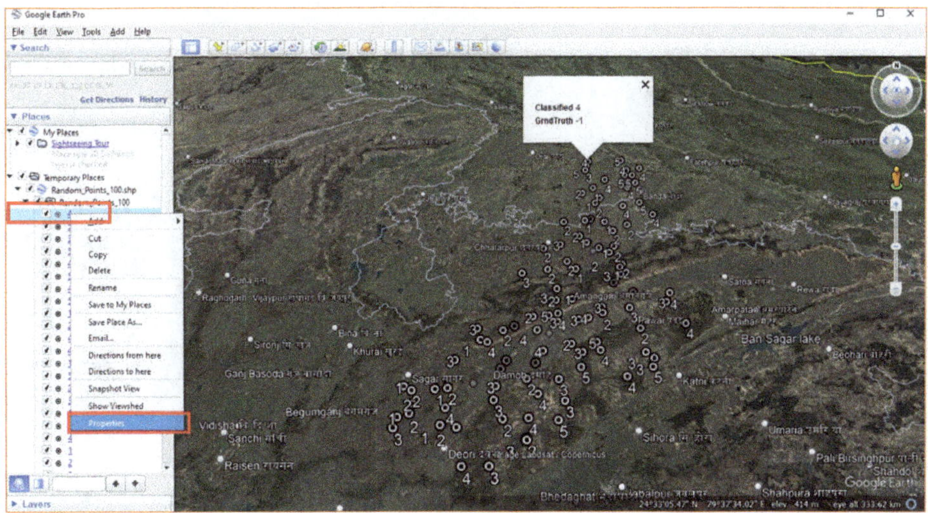

Fill the point number & it's class details in **Description** option and click on **OK** option.

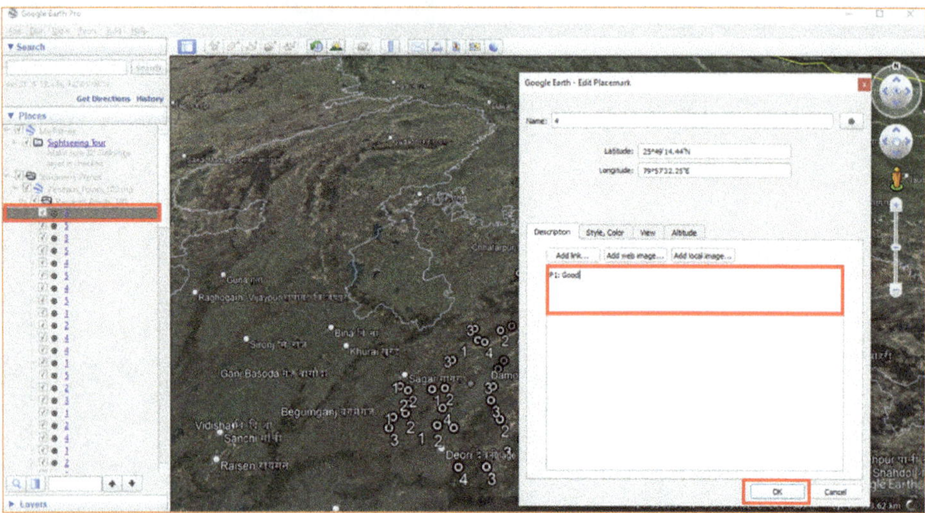

Point detail for Point 1 has been updated.

For verification, there is a strong need to update the information of all the points.

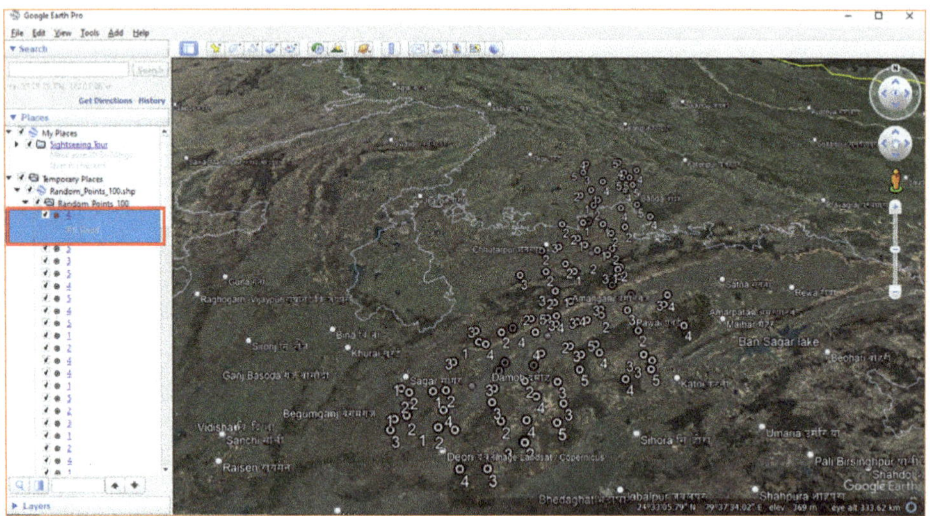

In this study, validation of the groundwater potential zones map of Ken River Basin has been done through the "Ground Water Prospect Study" map available on Bhujal-Bhuvan portal which provides the spatial information on well yield. This geospatial platform was developed by the CGWB and NRSC under the Rajiv Gandhi National Drinking Water Mission Project.

Open any browser and search Bhujal-Bhuvan Portal or click on the given link:

(https://bhuvan-app1.nrsc.gov.in/gwis/gwis.php)

Click on **Tools** and go to **Add Layer** option.

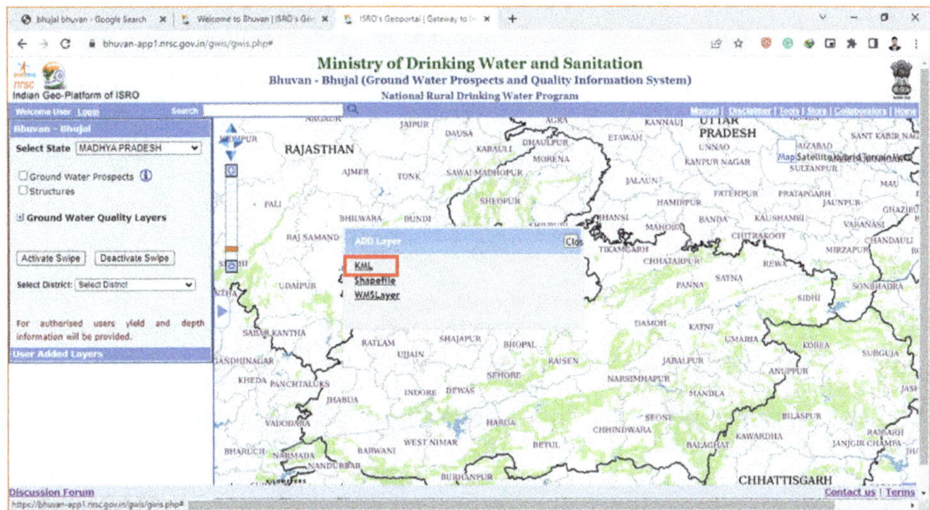

Click on **KML** option.

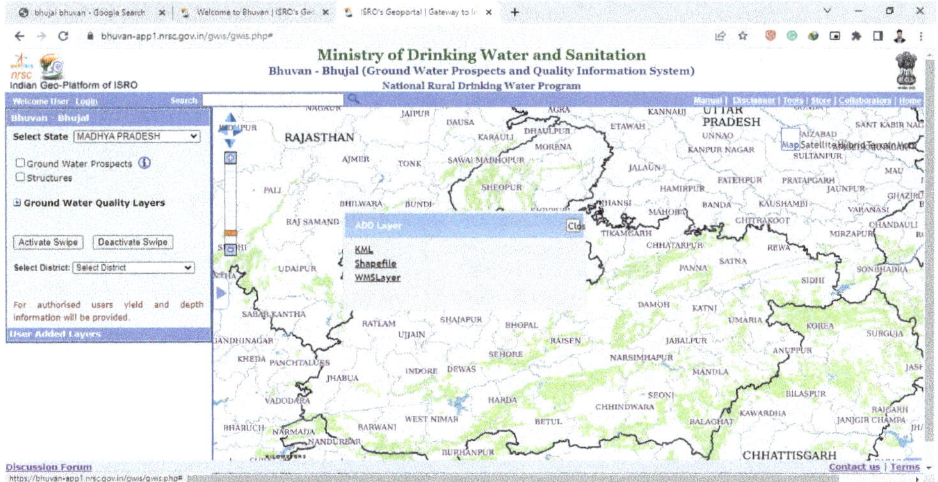

Browse the KML file name **RP.kml and click on Open** option.

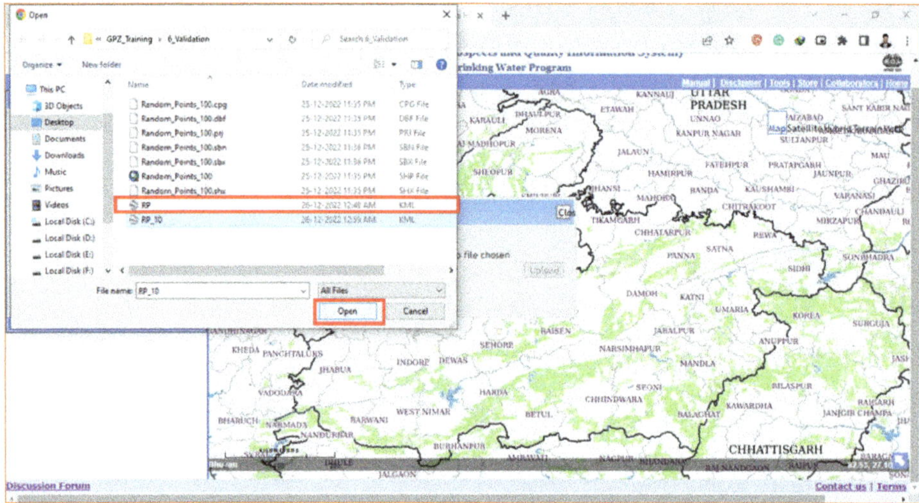

Accuracy Assessment Points have been imported in Bhujal-Bhuvan Portal.

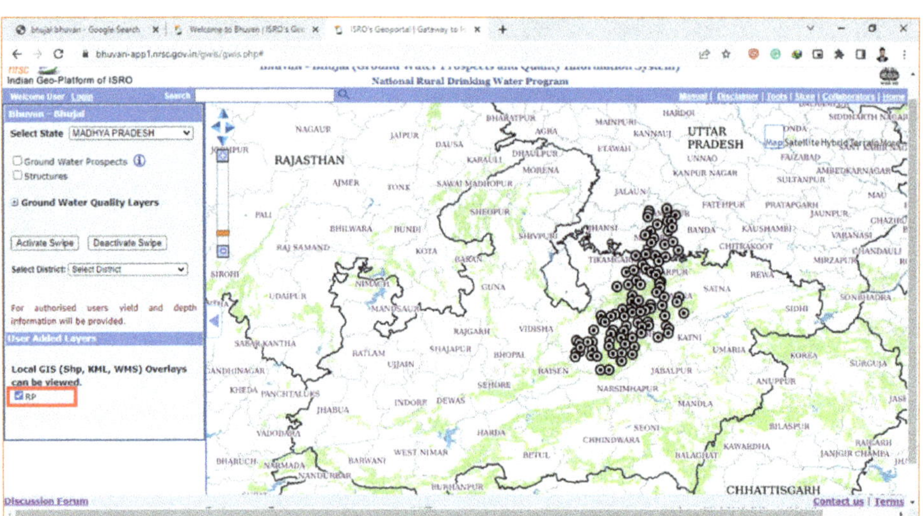

Select the **State** option and check the **Ground Water Prospect** option.

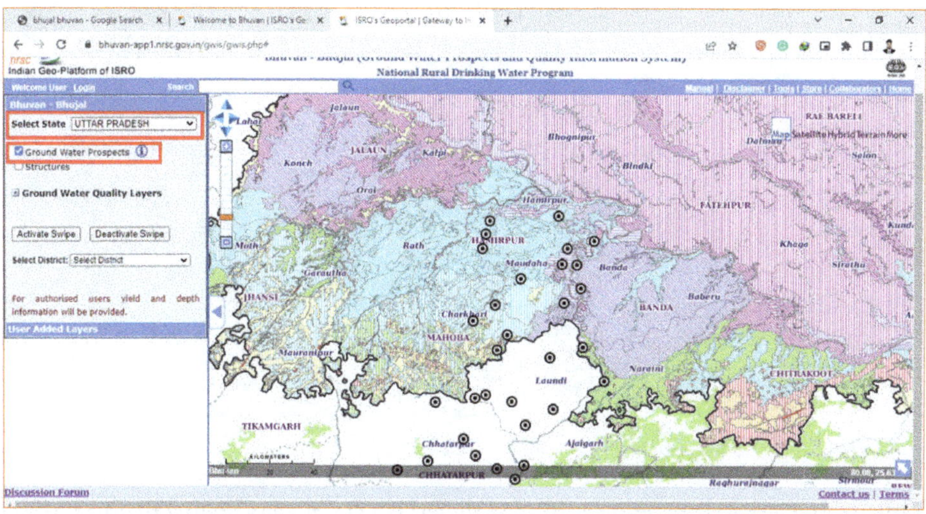

In Bhujal-Bhuvan portal, Madhya Pradesh/ Uttar Pradesh selected in state option and checked the 'Ground Water Prospect' option. Afterward, clicked on each point and zoom-in it. Then, identified the *colour* code and *lining* pattern of existing polygon. Classification of well yield ranges was given in the User manual of "Groundwater Prospect Mapping" (NRSA, 2011). The available yield ranges of well are classified into six classes from 10-50 LPM to > 800 LPM. These ranges of well yield are regrouped into five categories from < 50 LPM to > 400 LPM.

The well yield ranges extracted for all the random points, by clicking on each point manually using zoom-in & zoom-out option in Bhujal-Bhuvan portal.

This information has been obtained based on colour code and lining pattern of existing polygon below the particular point.

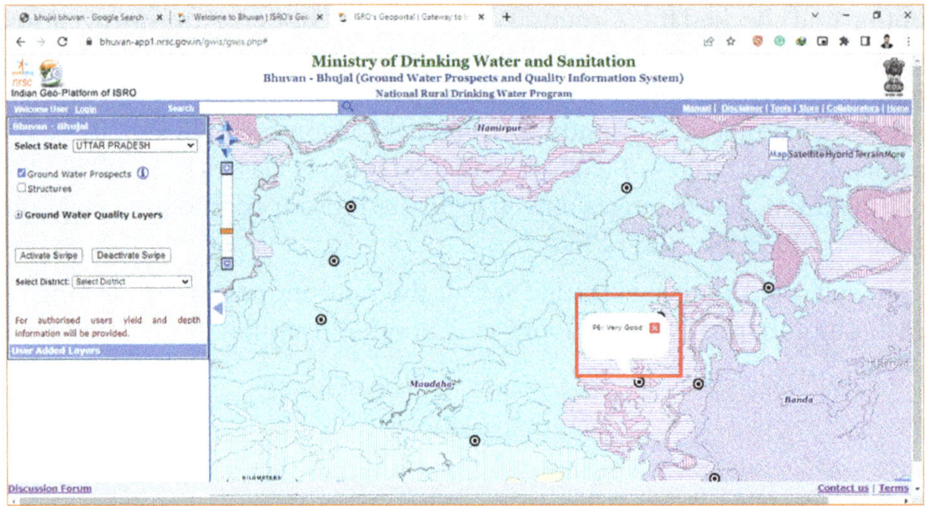

Both the data, well yield classes and GPZ classes were categorized into five categories like very good to very poor (Table 3).

Color Coding given in User manual of Groundwater Prospect

GROUND WATER PROSPECTS INFORMATION

YIELD RANGE OF WELLS	COLOUR CODE	DEPTH RANGE OF WELLS: SHALLOW < 30 METERS	MODERATE 30-80 METERS	DEEP > 80 METERS
> 800 LPM	Violet			
400-800 LPM	Indigo			
200-400 LPM	Blue			
100-200 LPM	Green			
50-100 LPM	Yellow			
10-50 LPM	Orange			
Prospects limited to valley portions only (Hills, Plateaus etc.)	Red			
Run-off zone/ Barrier for G.W. movement		(Inselberg / Ridge / Dyke etc.)		

Table 3: Classification of well yield

Yield Range of Wells	Class
> 400 LPM	Very Good
200 - 400 LPM	Good
100 - 200 LPM	Moderate
50 - 100 LPM	Poor
< 50 LPM	Very Poor

After the extraction of GroundTruth information from the portal, open the Arc Map software. Browse the validation folder, select the file **Random_Points_100.shp**, and click on the **Add** option.

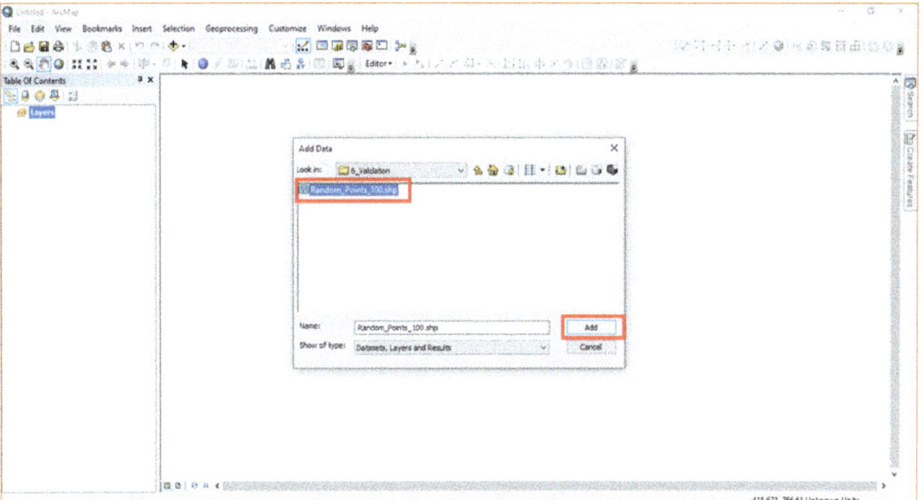

Right-Click on file name and click on Open Attribute Table option.

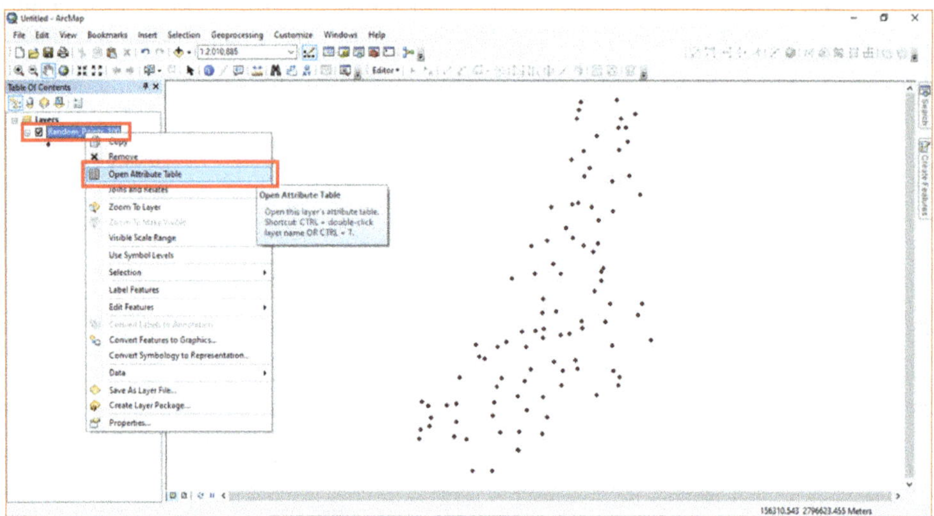

Click on **Switch Selection** option.

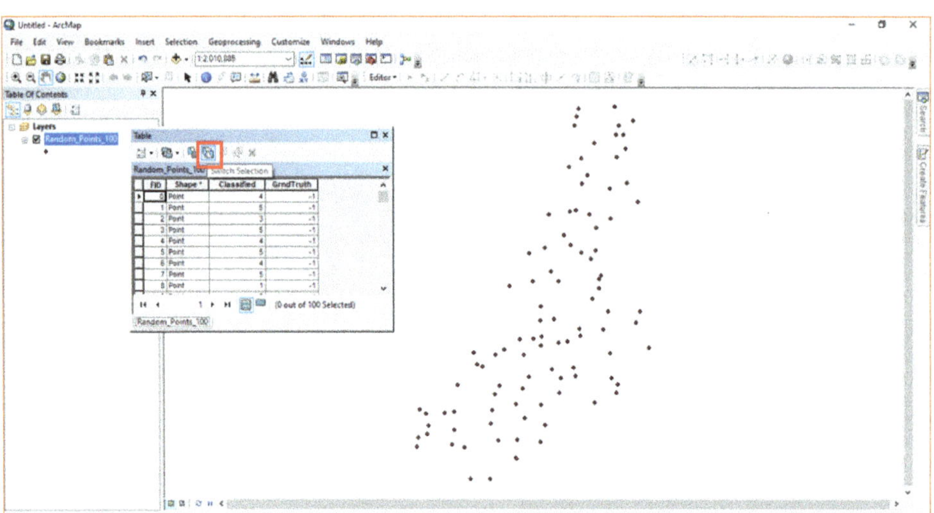

All the Points are selected.

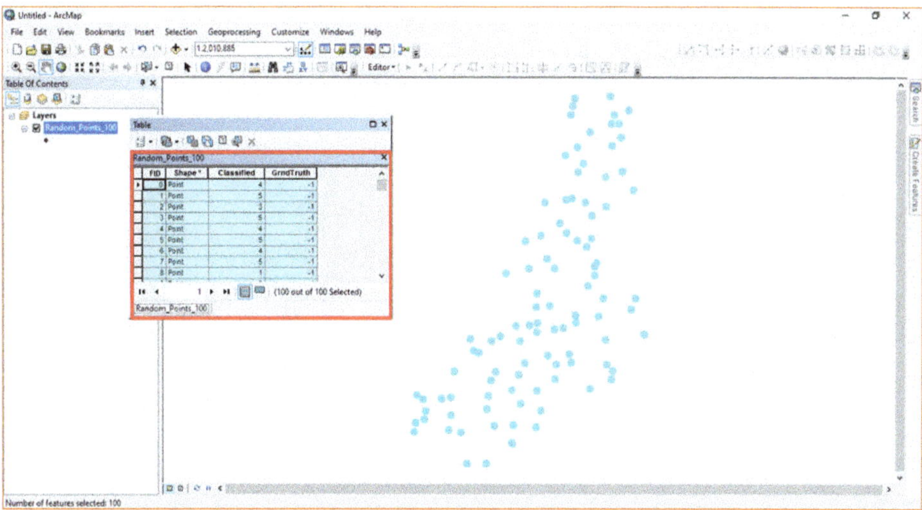

Right-Click on the corner of any row and click on **Copy Selected** option.

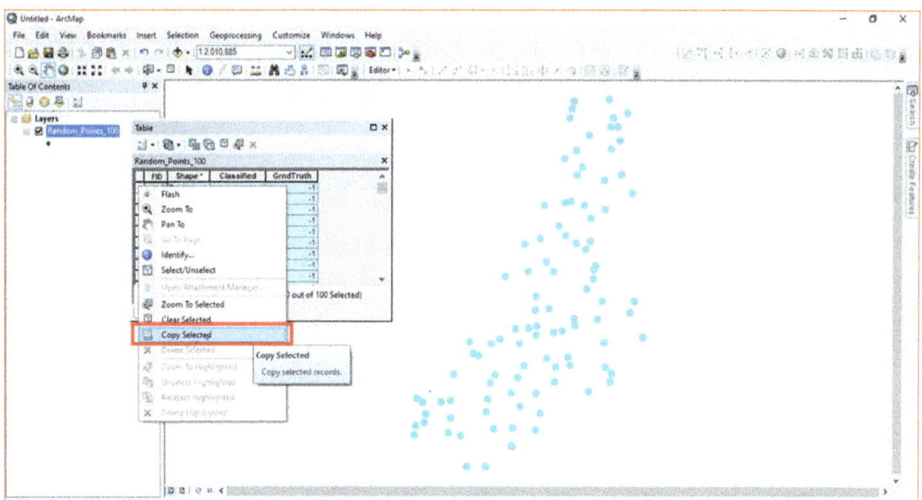

Paste (Ctrl + V) in the Excel sheet.

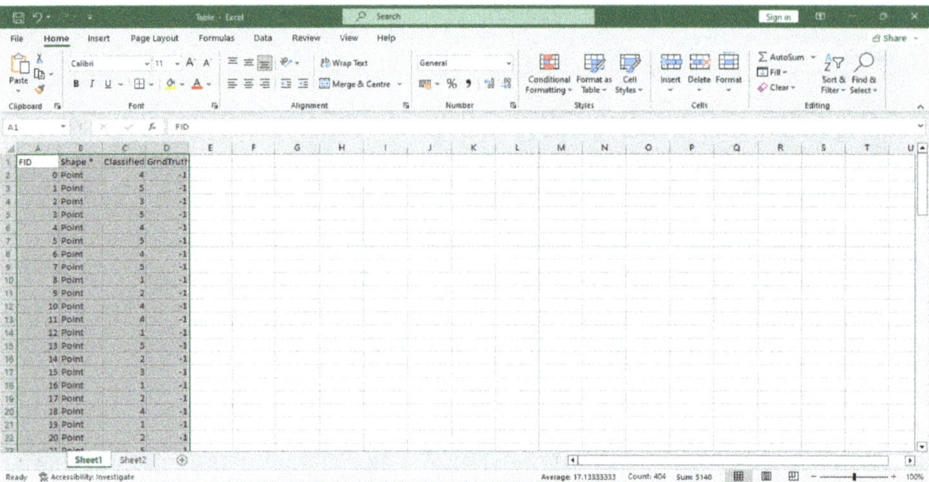

Add Classified column and fill the class name for all points. Remove the **-1 value** from the GroundTruth column. Fill the information on GroundTruth column.

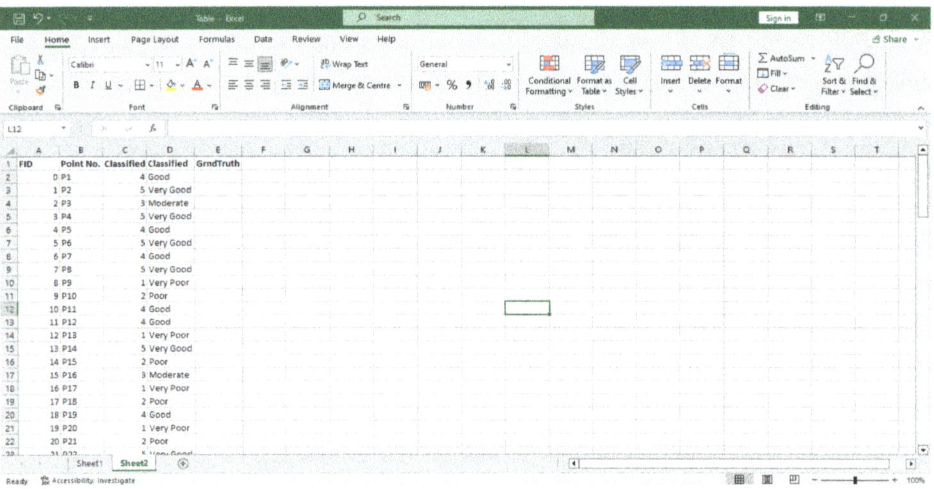

Add the column **Agreement/ Statement.**

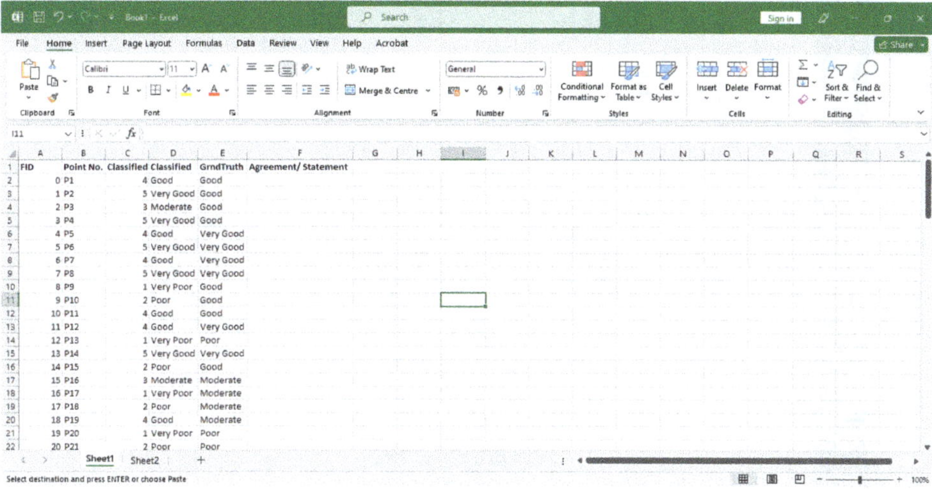

FID	Point No.	Classified	Classified	GrndTruth	Agreement/ Statement
0	P1	4	Good	Good	
1	P2	5	Very Good	Good	
2	P3	3	Moderate	Good	
3	P4	5	Very Good	Good	
4	P5	4	Good	Very Good	
5	P6	5	Very Good	Very Good	
6	P7	4	Good	Very Good	
7	P8	5	Very Good	Very Good	
8	P9	1	Very Poor	Good	
9	P10	2	Poor	Good	
10	P11	4	Good	Good	
11	P12	4	Good	Very Good	
12	P13	1	Very Poor	Poor	
13	P14	5	Very Good	Very Good	
14	P15	2	Poor	Good	
15	P16	3	Moderate	Moderate	
16	P17	1	Very Poor	Moderate	
17	P18	2	Poor	Moderate	
18	P19	4	Good	Moderate	
19	P20	1	Very Poor	Poor	
20	P21	2	Poor	Poor	

The appropriate dataset for validation of groundwater potential zones map is well yield of an area. The Bhujal-Bhuvan portal provides easy access to get spatial information on well yield. In Agreement Scheme approach, cross validation between GPZ classes and well yield ranges were done based on statements given in Table 4.

The Random Points containing info of GPZ class has been compared with the Well Yield ranges (Grundtruth) and prepared a table having Point no., Classified, GroundTruth, and agreements/ Statement.

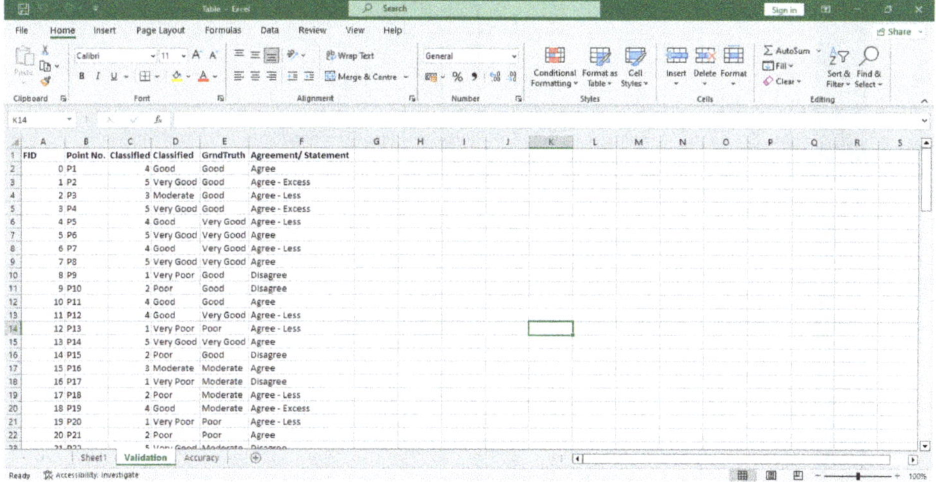

FID	Point No.	Classified	Classified	GrndTruth	Agreement/ Statement
0	P1	4	Good	Good	Agree
1	P2	5	Very Good	Good	Agree - Excess
2	P3	3	Moderate	Good	Agree - Less
3	P4	5	Very Good	Good	Agree - Excess
4	P5	4	Good	Very Good	Agree - Less
5	P6	5	Very Good	Very Good	Agree
6	P7	4	Good	Very Good	Agree - Less
7	P8	5	Very Good	Very Good	Agree
8	P9	1	Very Poor	Good	Disagree
9	P10	2	Poor	Good	Disagree
10	P11	4	Good	Good	Agree
11	P12	4	Good	Very Good	Agree - Less
12	P13	1	Very Poor	Poor	Agree - Less
13	P14	5	Very Good	Very Good	Agree
14	P15	2	Poor	Good	Disagree
15	P16	3	Moderate	Moderate	Agree
16	P17	1	Very Poor	Moderate	Disagree
17	P18	2	Poor	Moderate	Agree - Less
18	P19	4	Good	Moderate	Agree - Excess
19	P20	1	Very Poor	Poor	Agree - Less
20	P21	2	Poor	Poor	Agree

Table 4: Agreement for cross validation between GPZ classes and well yield ranges

SN	Well Yield Class	GPZ Class	Agreement	SN	Well Yield Class	GPZ Class	Agreement
1.	Very Good	Very Good	Agree	14.	Moderate	Poor	Agree - Less
2.	Very Good	Good	Agree - Less	15.	Moderate	Very Poor	Disagree
3.	Very Good	Moderate	Disagree	16.	Poor	Very Good	Disagree
4.	Very Good	Poor	Disagree	17.	Poor	Good	Disagree
5.	Very Good	Very Poor	Disagree	18.	Poor	Moderate	Agree - Excess
6..	Good	Very Good	Agree - Excess	19.	Poor	Poor	Agree
7.	Good	Good	Agree	20.	Poor	Very Poor	Agree - Less
8.	Good	Moderate	Agree - Less	21.	Very Poor	Very Good	Disagree
9.	Good	Poor	Disagree	22.	Very Poor	Good	Disagree
10.	Good	Very Poor	Disagree	23.	Very Poor	Moderate	Disagree
11.	Moderate	Very Good	Disagree	24.	Very Poor	Poor	Agree - Excess
12.	Moderate	Good	Agree - Excess	25.	Very Poor	Very Poor	Agree
13.	Moderate	Moderate	Agree				

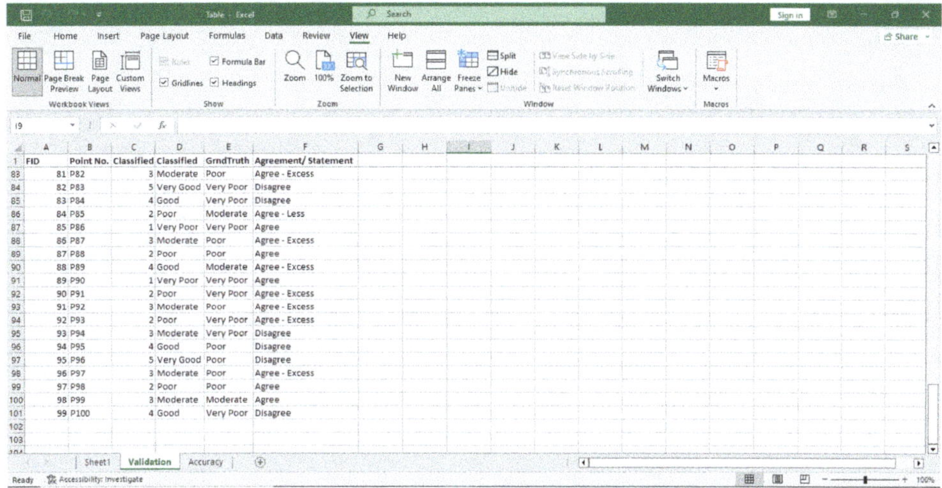

FID	Point No.	Classified	Classified	GrndTruth	Agreement/ Statement
81	P82	3	Moderate	Poor	Agree - Excess
82	P83	5	Very Good	Very Poor	Disagree
83	P84	4	Good	Very Poor	Disagree
84	P85	2	Poor	Moderate	Agree - Less
85	P86	1	Very Poor	Very Poor	Agree
86	P87	3	Moderate	Poor	Agree - Excess
87	P88	2	Poor	Poor	Agree
88	P89	4	Good	Moderate	Agree - Excess
89	P90	1	Very Poor	Very Poor	Agree
90	P91	2	Poor	Very Poor	Agree - Excess
91	P92	3	Moderate	Poor	Agree - Excess
92	P93	2	Poor	Very Poor	Agree - Excess
93	P94	3	Moderate	Very Poor	Disagree
94	P95	4	Good	Poor	Disagree
95	P96	5	Very Good	Poor	Disagree
96	P97	3	Moderate	Poor	Agree - Excess
97	P98	2	Poor	Poor	Agree
98	P99	3	Moderate	Moderate	Agree
99	P100	4	Good	Very Poor	Disagree

Overall accuracy has been estimated based on "agree condition (agree, agree-less, and agree-excess)" and "disagree" statement between well yield class and GPZ classes.

Accuracy (%) = No. of points in agree condition / total no. of points x 100

Accuracy = 80 / 100 x 100

Accuracy = 80 %

The validation accuracy can be classified into the following categories: 0.5 - 0.6 (poor) 0.6 - 0.7 (average); 0.7 - 0.8 (good); 0.8 - 0.9 (very good); and 0.9 – 1.0 (excellent) (Hosmer DW and Lemeshow SL, 2000).

The overall accuracy of Groundwater Potential Zones Map of Ken Basin has been estimated 80% which indicates the very good validation accuracy.

www.ingramcontent.com/pod-product-compliance
Ingram Content Group UK Ltd.
Pitfield, Milton Keynes, MK11 3LW, UK
UKHW021011290726
14059UKWH00001BA/75